3rd International Conference on Mechanical Manufacturing and Industrial Engineering 2020

IOP Conference Series: Materials Science and Engineering
Volume 947

Tokyo, Japan
26 - 28 August 2020

ISBN: 978-1-7138-3169-3
ISSN: 1757-8981

Printed from e-media with permission by:

Curran Associates, Inc.
57 Morehouse Lane
Red Hook, NY 12571

Some format issues inherent in the e-media version may also appear in this print version.

Institute of Physics
Dirac House, Temple Back
Bristol BS1 6BE UK

Phone: 44 1 17 929 7481
Fax: 44 1 17 920 0979

techtracking@iop.org

Additional copies of this publication are available from:

Curran Associates, Inc.
57 Morehouse Lane
Red Hook, NY 12571 USA
Phone: 845-758-0400
Fax: 845-758-2633
Email: curran@proceedings.com
Web: www.proceedings.com

TABLE OF CONTENTS

CONDENSED MATTER PHYSICS

Author Index

IOP Conf. Series: Materials Science and Engineering **947** (2020) 011001 doi:10.1088/1757-899X/947/1/011001

PREFACE

It is with great pleasure that I introduce the proceedings of the 2020 3rd International Conference on Mechanical Manufacturing and Industrial Engineering (MMIE 2020) which was held online recently at Hosei University, Tokyo, Japan during 26-28, August 2020.

The conference program is consisted of keynote lectures and oral presentations given by renowned scholars working in the areas of mechanical, manufacturing and industrial engineering from all over the world. To enrich the quality of program, some selected papers from 4th International Conference on Materials and Intelligent Manufacturing (ICMIM 2020), 3rd International Conference on Nano Science & Technology (ICNST 2020) were invited to be presented. It covered recent trends and progress made in the field of mechanical, manufacturing and industrial engineering. Distinguished researchers from Japan and Australia were invited to deliver four 45-minute keynote lectures regarding the latest research discoveries in their respective areas of expertise. Also, five sessions' authors made oral presentations and each presentation was confined in 15 minutes, following by question and answer. One excellent presenter will be chosen among them, depending on the technical merit, presentation file, Language fluency, and application etc for the best paper award.

The conference was managed with Zoom. Authors shared the screen and attendees could discuss their work via online platform. Overall, the conference was held successfully through the platform. Our conference committee and authors were dedicated to support the success.

All accepted papers were subject to peer-review by conference committee and invited international reviewers. The papers were selected based on their quality and their relevance to the conference.

I would like to express special gratitude to members of the conference committee and organizers of the conference. I would also like to thank the reviewers for their valuable time and advice which helped in improving the quality of the papers selected for presentation at the conference and for publication in the proceedings. Finally I want to thank the authors, the members of the organizing committee, the reviewers, the chairpersons, sponsors, and all other conference participants for their support of MMIE2020.

Prof. Alan Lau
Swinburne University of Technology, Australia
August 28, 2020

IOP Conf. Series: Materials Science and Engineering **947** (2020) 011002 doi:10.1088/1757-899X/947/1/011002

Conference Committee Chairs

Prof. Hideaki Tsukamoto, Hosei University, Japan
Prof. Alan Lau, Swinburne University of Technology, Australia

Technical Program Committee Chairs

Prof. Lihui Lang, Beihang University, China
Prof. Young Hoon Rim, Semyung University, Korea

Conference Poster Chair

Prof. Tingkai Zhao, Northwestern Polytechnical University, China

Technical Program Committee Members

Prof. Jian Yang, Shanghai Jiao Tong University, China
Prof. Shen-Yung Lin, National Formosa University, Taiwan
Prof. Arnold C. Alguno, MSU-Iligan Institute of Technology, Philippines
Assoc. Prof. Xiang Li, University of Shanghai for Science and Technology, China
Assoc. Prof. Rey Y. Capangpangan, Caraga State University, Philippines
Assoc.Prof. Radhika Pathak, Priyadarshini Institute of Architecture and Design studies, India
Asst. Prof. Ana I. Almerich Chulia, Universitat Politècnica De València, Valencia, Spain
Asst. Prof. Ibetombi Soibam, NIT Manipur, India
Asst. Prof. R.Yuvakkumar, Alagappa University, India
Dr. Fatih Dokme, Sisecam A.S., Turkey
Dr. Hao Wang, Northeastern University, China
Dr. Hendra Suherman, Universitas Bung Hatta, Indonesia
Chye Lih Tan, Universiti Malaysia Perlis, Malaysia
Yanisa Laoong-u-thai, Burapha University, Thailand
Yu-Cheng Chen, EPS bio technology corp., Taiwan
Zahira Yaakob, Universiti Kebangsaan Malaysia, Malaysia

Peer review declaration

All papers published in this volume of *IOP Conference Series: Materials Science and Engineering* have been peer reviewed through processes administered by the Editors. Reviews were conducted by expert referees to the professional and scientific standards expected of a proceedings journal published by IOP Publishing.

- **Type of peer review: Peer Review.**
- **Conference submission management system: ICONF.ORG**
- **Number of submissions received: 26**
- **Number of submissions sent for review: 26**
- **Number of submissions accepted:12**
- **Acceptance Rate (Number of Submissions Accepted / Number of Submissions Received X 100): 46%**
- **Average number of reviews per paper: 2**
- **Total number of reviewers involved: 11**
- **Any additional info on review process:**

  Preliminary Review
  The papers passed the first review will be reviewed again from the following aspects: Originality, Innovation, Technical Soundness, Applicability, and Clarity of presentation and Relevance

- **Contact person for queries:**
  Ashley Liu
  mmie@academic.net

Rolling Planning Method for Logistic System for Controlling Inventory and Stock-out Under Unsteady Demand

Md. Mohibul Islam and Masahiro Arakawa
Department of Architecture, Civil Engineering and Industrial Management Engineering, Nagoya Institute of Technology, Aichi 466-8555, Japan.

Corresponding author's e-mail address: m.islam.346@stn.nitech.ac.jp

Abstract. In this paper, rolling planning method for logistics system is proposed for minimizing logistics cost, controlling inventory, and stock out parts due to unsteady demand for a particular planning period. Demand variability is the main challenge for any supply chain network. One strategy to tackle this challenge is to practice a make-to-order supply system. However, there is a risk to have delays on the supply, when the final customers generate the order because the product requires a specific production time and a lead time to arrive from the factory to the customer. For handling unsteady demand, the main problem is the generation of stock-out products or in opposite case, the generation of dead stock due to a large number of storage products. For solving these problems, rolling planning method is proposed. The aim of this method is to estimate the optimal production, delivery quantities and to hold satisfied inventory level for minimizing stock out parts over the planning period. The proposed model and solution approaches are implemented in a numerical example. The solution has shown that the proposed rolling planning model for logistics system handle unsteady demand successfully for controlling inventory and stock out parts over the given planning period.

Keywords: Unsteady Demand, Rolling Method, Logistics System, Stock- out, Safety Stock.

1. Introduction

Inventory control is a fundamental key in any supply chain system. It plays an important role on the profit optimization, improving customers' satisfaction by reducing the number of stock-out products. Demand is variable in real supply chain system. This variability can be easily solved by the following make-to-order policy but this policy is not suitable for all products. For example, consumer items are typical items. For handling these items, producers follow make-to-stock policy. Under this policy, producers, wholesalers, and retailers can stock many items for meeting customers' demands. However, a large inventory easily generates overstock while a small inventory generates stock out in supply chain network. So, the challenge is to simultaneously reduce the inventory and stock-out parts. In many conventional supply chains, production and distribution planning are treated separately. However, they are mutually related problems that must be tackled in an integrated way [1]. The main characteristics of the past studies are mentioned in Table 1 in summarised form. According to past studies, most of the researchers have tried to construct multistage logistics model considering deterministic demand [1-8], stochastic demand [9-12] and fuzzy demand [13-15]. However, these are not sufficient to cope with realistic problem because real customers' demands are unsteady in nature and nobody has developed logistics model based on unsteady demand. Hence, extensions are needed for the previous model to make the model more realistic.

Table 1: Summary of past studies

	Demand characteristics of existing model			
Authors	Deterministic Demand	Stochastic Demand	Fuzzy Demand	Unsteady Demand
Fahimnia, Luong and Marian [1]	O	X	X	X
Seyedhosseini and Ghoreyshi [2]	O	X	X	X
Wang and Cheng [3]	O	X	X	X
Han et al. [4]	O	X	X	X
Coccola et al. [5]	O	X	X	X
Altiparmak et al. [6]	O	X	X	X
Gan, Li and Si [7]	O	X	X	X
Su et al. [8]	O	X	X	X
Manupati et al. [9]	X	O	X	X
Aaron et al. [10]	X	O	X	X
Nourifar, R. et al. [11]	X	O	X	X
Zhang, Shang and Li [12]	X	O	X	X
Gharehyakheh and Moghaddam [13]	X	X	O	X
Sakalli [14]	X	X	O	X
Jamrus et al. [15]	X	X	O	X

Notation: "O" = Considered and "X" = Did not Considered.

In this study, rolling planning method for logistics system is proposed to control inventory and stock-out parts and to minimize logistic cost under unsteady demands in make-to-stock production. The proposed model has following characteristics: (i) Initial inventory at the store will be estimated based on average customers' demands and adding the safety inventory and initial inventory at the plant and DCs will be calculated based on the average demand and add the sum of all the variances of the stores times the safety factor "$\delta = 1.88$". (ii) During the demand consumption at the stores (when $t = 1$), the production volume of the plant will be equal to the initial inventory of that plant and after initial period when $t \geq 2$ the plant will produce the products based on the rolling planning method, which means that the production quantities at period "t" will be the sold quantities on all the stores in period "t-1". (iii) Inventory will be revised at the end of each period. (iv) Products will be delivered from plant to DCs and DCs to stores based on order and arrive according to lead time (LT). (v) LT between plant and DCs requires one period of time and LT between DCs and stores requires another one period of time (vi) Real customers' demands will be unsteady which follow normal distribution with a known mean and a standard deviation. (vii) If stock-out occurs at any store, it will be delivered next period.

2. Model

Figure 1 shows the time chart of the orders and the deliveries to each location. The dotted lines represent new orders requiring delivery and the solid lines denote the delivery of products. Here, the plant, DCs, and the stores have identical times at the end and the start of every period. Demand is generated at the retail shops between the start and the end of every period. A logistics plan is prepared when a new order for a delivery arrives at the DCs from the retail shops. Then, from the logistics plan, the number of products transported from DCs to shops and the number of products transported from the plant to DCs are determined. The shop requests new orders from the DCs after the end of the period. Products transported from DCs arrive at shop by the start of the next period according to the logistics plan. In addition, the inventory of DCs is examined, and the DCs request new orders from the plant. The plant transports products to the DCs according to the orders requested from the DCs. The products arrive at the DCs by the start of the next period. Therefore, products produced in the plant in the t-th period arrive at shops at the start of the (t+2)-th period. Figure 2 represents the two-stage delivery model with inventory representation and figure 3 shows the flow chart of rolling planning method. The mathematical model of this problem has 45 constraint functions including objective function among them a few of them are denoted as follows:

Indexes: i: Plant ($i = 1, 2, \ldots I$), j: Distribution centres, DCs ($j = 1, 2, \ldots J$), k: Clusters ($k = 1, 2, \ldots K$), t: Periods ($t = 1, 2, \ldots T$), s: Stores ($s = 1, 2, \ldots S$), p: Products ($p = 1, 2, \ldots P$).

Parameters: $\widehat{D}^{S}_{t,p,k}$: Real demand at period "t" of cluster "k" for product "p", $ca_{i,j}$: Delivery cost from plant "i" to DC "j", $cb_{j,k}$: Delivery cost from DC "j" to Clusters "k", CI^{D}_{j} : Inventory capacity at the DC "j", M: Large integer number.
Decision Variables: $P_{t,p,i}$: Production quantity of product "p" at plant "i" at the period of "t", $O^{D}_{t:t+1,p,i,j}$: Quantities delivered at the end of period "t" from plant and arrived before t+1 to DC "j" for product "p", $O^{S}_{t:t+1,p,j,k}$: Quantities delivered at the end of period "t" from DC and arrived before t+1 to Cluster "k" for product "p", $I^{S}_{t,p,k}$: Inventory in Clusters "k" at end of period "t" for product "p", $I^{D}_{t,p,j}$: Inventory in DC "j" at the end of period "t" of product "p", $I^{F}_{t,p,i}$: Inventory at the end of period "t" for plant "i" and product "p", $A_{p,t,k}$ Number of stock-out parts at period "t" for product "p".
Objective function: Minimize the logistics cost from Plant to DC and from DC to cluster for product "p" on period "t". The equation is shown as follow:

minimize: $\sum_{t=1}^{T}\sum_{p=1}^{P}\sum_{i=1}^{I}\sum_{j=1}^{J} O^{D}_{p,t:t+1,i,j}\, ca_{i,j} + \sum_{t=1}^{T}\sum_{p=1}^{P}\sum_{j=1}^{J}\sum_{k=1}^{K} O^{S}_{p,t:t+1,j,k}\, cb_{i,j}$ (1)

Initial inventory for each cluster is calculated by equation (2) which considers the average customer demands and adding the safety inventory.

$$I^{S}_{c,p,k} = \frac{\sum_{s=1}^{S}(D^{S}_{s,p} + \delta\sigma_{s,p}\sqrt{LT})}{K} \quad (2)$$

where, $D^{S}_{s,p}$: Average Demand at store "s" for product "p", $\sigma_{s,p}$ = Standard Deviation for "s" stores for product "p", δ = Safety factor, LT: Lead Time, for practical purposes the lead time is considered one unit of time for all the deliveries from the plant to DC and from DC to each store. For meeting the customer demands at each store, the distribution centers contain safety inventory for the supply of the products. The initial inventory for each DC is given by adding the sum of the average demand plus δ times the square root of the sum of all the variances of the stores and is given by the following formula:

$$I^{D}_{c,j,p} = \frac{(\sum_{s=1}^{S} D^{S}_{s,p}) + \delta \times \sqrt{\sum_{s=1}^{S}\sigma^{2}_{s,p}}}{J} \quad (3)$$

For the initial inventory at the manufacturing site, $I^{F}_{c,i,p}$ is given by equation (4) and the same values of the production quantities $P_{t,i,p}$ at the first period are defined by equation (5). The quantities are calculated based on the average demand and add the sum of all the variances of the stores times the safety factor.

$$I^{F}_{c,i,p} = \frac{(\sum_{s=1}^{S} D^{S}_{s,p}) + \delta \times \sqrt{\sum_{s=1}^{S}\sigma^{2}_{s,p}}}{I} \quad (4)$$

$$P_{t=1,i,p} = \frac{(\sum_{s=1}^{S} D^{S}_{s,p}) + \delta \times \sqrt{\sum_{s=1}^{S}\sigma^{2}_{s,p}}}{I} \quad (5)$$

Production quantity at period "t >1" follows a rolling planning schedule that is these quantities are defined based on the real customer's demand from period "t-1" as given in equation (6).

$$P_{t>1,p,i} = \sum_{k=1}^{K} \widehat{D}^{S}_{t-1,p,k} \ (\forall\ i, k) \quad (6)$$

Inventory at the end of the first period (t = 1) for each cluster of product "p" is calculated based on the initial inventory that is being consumed by the real demand and adding the stock out parts in that period as given in equation (7).

$$I^{S}_{1,p,k} = I^{S}_{c,p,k} - \widehat{D}^{S}_{1,p,k} + A_{1,p,k} \ (\forall p, k) \quad (7)$$

Inventory at the end of the period "t" when t > 1 for each cluster of product "p" is defined by the previous period inventory adding the arrival parts from the DC subtracting the real demand and considering the ordering of the stock out parts as given in equation (8). Equation (9) is the restriction of avoiding stock out parts considering the inventory and the delivered quantities on the previous term at each store. Inventory at the end of the first period (t= 1) for each DC is defined by the initial inventory subtracting the quantities to be delivered to each cluster for the second period as given in equation (10).

$$I^S_{t,p,k} = I^S_{t-1,p,k} + \sum_{j=1}^{J} O^S_{t-1:t,p,j,k} - \widehat{D}^S_{t,p,k} - A_{t-1,p,k} + A_{t,p,k} \; (\forall \text{p}, \text{k}, \text{t} \in \text{T}, \text{t} \geq 2) \tag{8}$$

$$I^S_{t-1,p,k} + \textstyle\sum_{j=1}^{J} O^S_{t-1:t,p,j,k} \geq \widehat{D}^S_{t,p,k} + A_{t-1,p,k} - A_{t,p,k} \; (\forall \text{p}, \text{k}, \text{t} \in \text{T}, \text{t} \geq 2) \tag{9}$$

$$I^D_{1,p,j} = I^D_{c,p,j} - \textstyle\sum_{k=1}^{K} O^S_{1:2,p,j,k} \quad (\forall \text{p}, \text{j}) \tag{10}$$

Inventory in the DC "j" at end of period "t >1" of product "p" is the result at the end of the period is the sum of the inventory at the beginning of the period and the received quantities from the plant, subtracting the quantities to be delivered to each cluster as given in equation (11).

$$I^D_{t,p,j} = I^D_{t-1,p,j} + \textstyle\sum_{i=1}^{I} O^D_{t-1:t,p,i,j} - \sum_{k=1}^{K} O^S_{t:t+1,p,j,k} \quad (\forall \text{p}, \text{j}, \text{t} \in \text{T}, \text{t} \geq 2) \tag{11}$$

Equation (12) is defined as the inventory at plant "i" at the end of the first period (t = 1). Equation (13) is defined as the inventory at plant "i" at period "t" of product "p" after the first period (t>1). The inventory is defined by sum of the previous inventory and the production quantity subtracting the quantities that will be delivered at the end of the period to each DC.

$$I^F_{1,p,i} = I^F_{c,p,i} + P_{1,p,i} - \sum_{j=1}^{J} O^D_{1:2,p,i,j} \; (\forall \text{p}, \text{i}) \tag{12}$$

$$I^F_{t,p,i} = I^F_{t-1,p,i} + P_{t,p,i} - \textstyle\sum_{j=1}^{J} O^D_{t:t+1,p,i,j} \; (\forall p, i, \text{t} \in \text{T}, \text{t} \geq 2) \tag{13}$$

Equation (14) and (15) denote constraints related to delivery quantity from DC "j" to Cluster "k" and from plant "i" to DC "j" when t >1.

$$I^D_{t-1,p,j} + \textstyle\sum_{i=1}^{I} O^D_{t-1:t,p,i,j} \geq \sum_{k=1}^{K} O^S_{t:t+1,p,j,k} \; (\forall \text{p}, \text{j}, \text{t} \in \text{T}, \text{t} \geq 2) \tag{14}$$

$$P_{t,p,i} + I^F_{t-1,p,i} \geq \textstyle\sum_{j=1}^{J} O^D_{t:t+1,p,i,j} \; (\forall \text{p}, \text{i}, \text{t} \in \text{T}, \text{t} \geq 2) \tag{15}$$

Equations (16) and (17) show the inventory capacity constraint at DC "j" of the 1st period and when t ≥ 2.

$$\textstyle\sum_{p=1}^{P} I^D_{c,p,j} + \sum_{p=1}^{P}\sum_{i=1}^{I} O^D_{0:1,p,i,j} \leq CI^D_j \; (\forall \text{p}, \text{t} = 1, j) \tag{16}$$

$$\textstyle\sum_{p=1}^{P} I^D_{t-1,p,j} + \sum_{p=1}^{P}\sum_{i=1}^{I} O^D_{t-1:t,p,i,j} \leq CI^D_j \; (\forall j, \text{t} \in T, \text{t} \geq 2) \tag{17}$$

Equation (18) shows the total delivery quantity constraint from "t" period at DC "j" to Cluster "k" for product "p". If there is a stock out event at any store, the binary variable $B^S_{t,p}$ will be 1 and will not allow to delivery parts. Same types of constraints are introduced first period and when t ≥ 2.

$$\textstyle\sum_{k=1}^{K}(I^S_{c,p,k} - I^S_{t,p,k} + A_{t,p,k}) - B^S_{t,p} * M \leq \sum_{j=1}^{J}\sum_{k=1}^{K} O^S_{t:t+1,p,j,k} \; (\forall \text{p}, \text{t}, \text{k}) \tag{18}$$

3. Numerical Experiment

In order to evaluate the performance of the proposed model, a numerical experiment is performed. Assumed, I = 1, J = 2, K = 3, S = 20, T = 5, LT =1, and P = 2 (A and B), unit logistic costs from plant to DC1 and DC2 are 14 and 15 respectively and from DC1 to K1, K2, K3 are 15, 5,150 and DC2 to K1, K2, K3 are 150, 15 and 15 respectively. Random data based on normal distribution are generated to get as real demand. Mean demand and standard deviation of products A and B are predetermined as 120, 80 and 20, 20 respectively for 20 stores. By using this data, initial inventories for clusters, DCs, and a plant are estimated as shown in Table 2. Unsteady customers' demands are given in Table 3.

4. Results and Discussion

Initially, products A and B are produced at plant considering the initial inventory of a plant, and after the initial period when t > 1, products A and B are produced in the plant according to the rolling planning method which is given in Table 4. At the end of every period, products are delivered from plant to DCs based on order repeatedly as shown in Table 5, and then ending inventory at the plant is also calculated periodically which is given in Table 6. Again, in the same way, initially and subsequently, products are delivered from DCs to clusters based on order as shown in Table 7 and then ending inventory at the DCs is calculated periodically which is given in Table 8. In terms of clusters, customers' unsteady demands for every period which are given in Table 3 are fulfilled by clusters from their stock items, and then ending inventory of clusters for every period is revised which is given in Table 9. Finally, according to equation (1), the total logistic cost is calculated by this proposed model as shown in Table 10. It is observed that by evaluating the proposed model there is no stock-out situation as shown in Table 11. For testing the feasibility of the proposed model, the model is performed by setting another safety factor $\delta = 0$ and get the total logistic cost as given in Table 10 which is slightly lower than the previous result. But, under this condition, stock-out is occurred as shown in Table 11. It's proved that the proposed model is generating the results satisfactorily. It is mentioned that if safety factor $\delta = 0$ is considered, service level will be only 50% which means that there is a possibility to stock out about 50% which is not convenient to hold customer satisfaction while if safety factor $\delta = 1.88$ is considered, service level will be 97% which is very convenient to hold customer satisfaction. Therefore, by evaluating inventory level and stock-out situation, we consider that the determined logistic cost, which is 592822, is acceptable in this model.

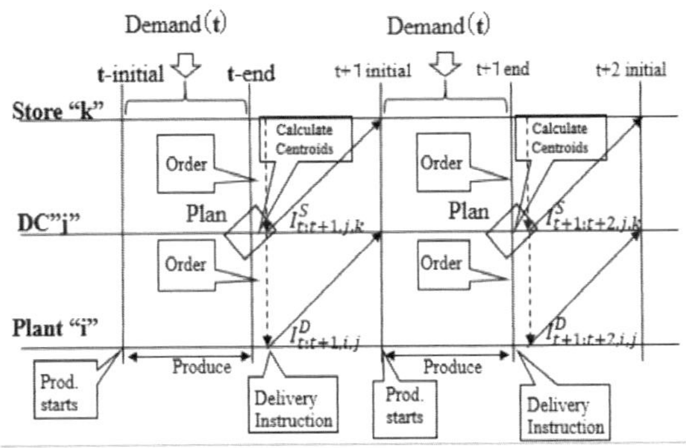

Figure 1. Time Chart

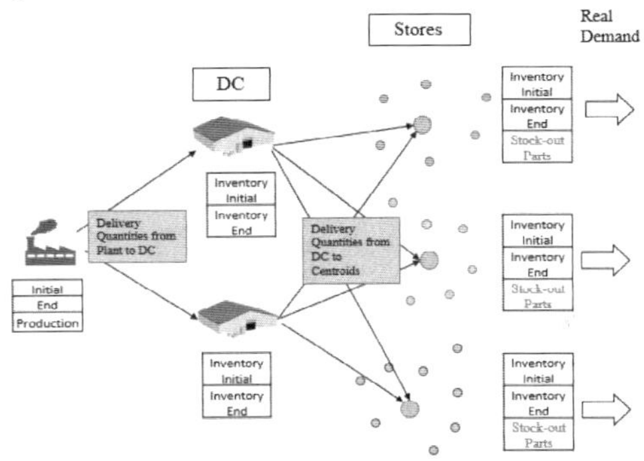

Figure 2. Two-Stage Inventory Representation

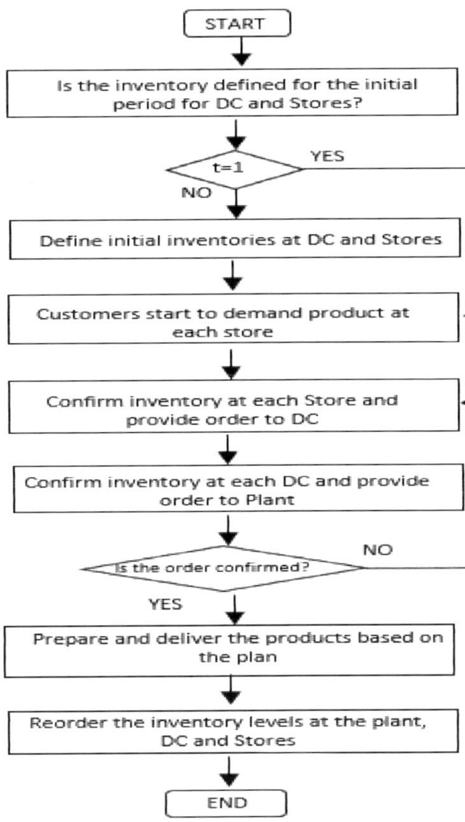

Figure 3. Flow Chart of Rolling Planning Method

5. Conclusion

In this study, a rolling planning method for logistics system is proposed that simultaneously reduces the total inventory, logistics cost, and stock-out. The proposed model considers the following controls: production, standard inventory for meeting unsteady demands and the distribution of products from plant to DCs and from DCs to shops to minimize discrepancy between requirements and supply. In this study, stock-out and over-stock are reduced in make-to-stock production system which is very essential to control inventory in supply chain network. In future studies, this model can be extended by integrating delivery routes for finding the optimal logistics cost in a large-scale problem.

Table 2. Initial inventory at clusters, DCs and plant.

	$\delta = 0$		$\delta = 1.88$	
	A	**B**	**A**	**B**
K1, K2, K3	800	534	1051	784
DC1, DC2	1200	800	1285	885
I	2400	1600	2570	1770

Table 3. Real unsteady customers' demands.

	K1		K2		K3	
Periods	A	B	A	B	A	B
T1	851	502	929	552	687	490
T2	902	564	764	598	711	450
T3	807	624	874	597	659	477
T4	807	591	782	638	724	533
T5	785	581	842	582	747	561

Table 4. Production volume at plant.

Products	T1	T2	T3	T4	T5
A	2570	2467	2377	2340	2313
B	1770	1544	1612	1698	1762

Table 5: Delivery volume from plant to DCs.

DCs	Products	T1	T2	T3	T4	T5
DC1	A	1666	1681	1589	2912	0
	B	1954	1544	114	2048	0
DC2	A	1893	0	659	1471	0
	B	1586	0	0	1094	0

Table 6. Inventory at plant at the end of every period.

Products	T1	T2	T3	T4	T5
A	1581	2367	2496	453	2766
B	0	0	1498	54	1816

Table 7. Delivery volume from DCs to clusters.

		K1		K2		K3	
DCs	Periods	A	B	A	B	A	B
DC1	T1	851	502	434	383	0	0
	T2	902	564	764	598	0	0
	T3	807	624	874	597	0	0
	T4	807	591	782	638	0	0
	T5	785	581	842	582	0	0
DC2	T1	0	0	495	169	687	490
	T2	0	0	0	0	711	450
	T3	0	0	0	0	659	477
	T4	0	0	0	0	724	533
	T5	0	0	0	0	747	561

IOP Conf. Series: Materials Science and Engineering **947** (2020) 012001 doi:10.1088/1757-899X/947/1/012001

Table 8. Inventory at DCs at the end of every period.

	DC1		DC2	
Periods	A	B	A	B
T1	0	0	103	226
T2	0	792	1285	1362
T3	0	1115	626	885
T4	0	0	561	352
T5	1285	885	1285	885

Table 9. Inventory at clusters at the end of every period.

	K1		K2		K3	
Periods	A	B	A	B	A	B
T1	200	282	122	232	364	294
T2	149	220	287	186	340	334
T3	244	160	177	187	392	307
T4	244	193	269	146	327	251
T5	266	203	209	202	304	223

Table 10. Total logistic cost.

	Safety factors	
	$\delta = 0$	$\delta = 1.88$
Total Logistic Cost	582618	592822

Table 11. Stock-out at clusters at the end of every period.

	$\delta = 0$		$\delta = 1.88$	
Periods	A	B	A	B
T1	180	18	0	0
T2	133	94	0	0
T3	81	153	0	0
T4	7	258	0	0
T5	42	382	0	0

References:

[1] Fahimnia B, Luong L and Marian R. An integrated model for the optimization of a two-echelon supply network. *JAMME*, 2008, vol. **31**, issue 2.

[2] Seyedhosseini SM, and Ghoreyshi SM. An integrated model for production and distribution planning of perishable products with inventory and routing considerations. Math. Prob. Engineering, vol. 2014.

[3] Wang X, and Cheng TCE. Logistics scheduling to minimize inventory and transport costs. Int. J. Production Economics.2009, vol.121, issue 1, pp. 266-273.

[4] Hang JH, Lee JY and Kim YD. Production planning in a two-level supply chain for production-time dependent products with dynamic demands. Comp.& Ind. Engineering, 2019, vol. 135, pp. 1-9.

[5] Coccola ME, Carlos A, Rodolfo MG, and Dondob. A two-stage procedure for efficiently solving the integrated problem of production, inventory, and distribution of industrial products. Com. and Chem. Engineering, 2019, vol. 134.

[6] Altiparmak MFGen LL and Paksoy T. A genetic algorithm approach for multi-objective optimization of supply chain networks, Com. & Ind. Engineering, vol. 51, no. 1, pp. 196–215.

[7] Gan M, Li Z, and Si C. On the Transformation Mechanism for Formulating a Multiproduct Two-Layer Supply Chain Network Design Problem as a Network Flow Model. 2014, Math. Prob. Engineering, 2014, vol. 2014.

[8] Su Y, Qin J, Yang P and Jiang Q. A Supply Chain-Logistics Super-Network Equilibrium Model for Urban Logistics Facility Network Optimization. 2019, Math. Prob. Engineering, vol. 2019.

[9] Manupati VK, Jedidah SJ, Gupta S, Bhandari A and Ramkumar M. Optimization of a multi-echelon sustainable production-distribution supply chain system with lead time consideration under carbon emission policies 2018 Com. Ind. Engg. vol.135, pp. 1312-23.

[10] Aaron GC, Benitez EO, Miranda PA, Loaiza REP and Rosas JHA. 2018, Design of a Logistics Nonlinear System for a Complex, Multi echelon, Supply Chain Network with Uncertain Demands. 2018, Complexity, vol. 2018.

[11] Nourifar R, Mahdavi I, Amiri NM and Payda MM. Optimizing decentralized production–distribution planning problem in a multi-period supply chain network under uncertainty. 2017, J Ind. Eng. Int, vol. 14, pp.367–82.

[12] Zhang G, Shang J, and Li W. Production, Manufacturing and Logistics, Collaborative production planning of supply chain under price and demand uncertainty, 2011, EJOR, vol. 215, issue 3, pp. 590-603.

[13] Gharehyakheh A, and Moghaddam RT. A fuzzy solution approach for a multi-objective integrated production -distribution model with multi products and multi periods under uncertainty 2012 Mgt. Sc. L. vol. 2, pp. 2425-34.

[14] Sakalli US. Optimization of Production-Distribution Problem in Supply Chain Management under Stochastic and Fuzzy Uncertainties. 2017, Math. Prob. Engineering, vol. 2017.

[15] Jamrus T, Chien CF, Gen M and Sethanan K. Multistage production distribution under uncertain demands with integrated discrete particle swarm optimization and extended priority-based hybrid genetic algorithm 2015 Fuzzy Opt. Dec. M. vol. 14, pp. 265–87.

Development of information system to generate assembly order of parts considering work difficulty

Masahiro Arakawa[1], Souta Yokoi[1], Hayata Kobayashi[1], Sayaka Kurihara[2], and Kensuke Ishida[1]

[1] Department of Industrial Management Engineering, Nagoya Institute of Technology, Gokiso, Showa, Nagoya, 466-8555, Aichi, Japan
[2] Brother Industries, Ltd., Naeshio, Mizuho, Nagoya, 467-8561, Aichi, Japan

m.arakawa@nitech.ac.jp

Abstract. In this study, we develop an information system to generate the assembly order of parts to reduce work difficulty during parts assembly. Several types of tables are proposed to store the relationship of connection and contact between parts and characteristics related to the structure between parts causing work difficulty. In addition, a mathematical model is constructed to generate the assembly order of parts to reduce work difficulty. The effectiveness of the developed system is evaluated based on numerical and assembly experiments using an actual product.

1. Introduction

In many assembly processes of parts, multiprocess holding works are introduced because their flexibility allows complex works to be performed in a short time. Therefore, even in high-mix low-volume productions, production conditions are assessed and works are performed in a short time; therefore, workers must be able to flexibly manipulate a plurality of parts successively. In the mass production or continuous production of the same product, the work is repeated; therefore, learning and jigs can be easily introduced for productivity enhancement. However, when high-mix low-volume production or single-product manufacturing is enhanced, a process line tends to be constructed through the assignment of multiple handling works to single workers [1]. Although different works and parts are allocated to the process line for processing different products, similar products and works are included [2].

However, in the current manufacturing industry, several problems exist, such as difficulty in reducing workers, increasing wages, adjustment of number of workers owing to fluctuations in demand, and management of worker skills. Hence, to measure the productivity enhancement of parts assembly, we developed a method to determine the order of parts assembly that facilitates a reduction in work operation time. If we can seamlessly determine the assembly order of parts using information of Computer Aided Design (CAD) software for product design, then both product and work designs can be performed simultaneously, and the period of work design can be reduced. In addition, if an information system to estimate the operation time of a work process is developed, then works can be managed easily. To realize this mechanism, the following processes must be combined: a process for generating information to determine the assembly sequence from CAD; a process for calculating the sequence to assemble the parts; and a process for estimating the operation time.

In this study, an information system including the mechanism combining the processes above was developed, and the effectiveness of the system is demonstrated through a case study.

IOP Conf. Series: Materials Science and Engineering **947** (2020) 012002 doi:10.1088/1757-899X/947/1/012002

2. Evaluation method of work difficulty by product structure

The purpose of this study is to construct an information system combining product design and work design. We assumed that work difficulty is evaluated from the structure of parts that constitute a product. We focus on multiple factors that cause work difficulty to generate from the part structure and evaluate the difficulty by combining the evaluation values of factors. Parts assembly can be expressed as repeating the work of "assembling one part into a semi-finished product" and the assembly of a single part is regarded as one work unit. Subsequently, factors of work difficulty that affect the structure of parts appearing in one work unit are evaluated. We assume that this one work unit is composed of the following tasks:

(w1) picking and moving a single part,

(w2) assembling a single part into a semi-finished product.

Tasks (w1) and (w2), are used to calculate the operation time considering the difficulty using the method time measurement (MTM) method.

We categorize the factors causing the work difficulty into parts to assemble, semi-finished products, and support parts, such as bolts and nuts. Table 1 shows a list of the factors, which are used to evaluate the difficulty from the parts structure [3].

Table 1. Factors affecting operation difficulty in assembly.

Objects	Categories	Factors
Parts	Positioning	(1) Direction or side to assemble a single part
		(2) Size and weight
	Method to assemble a part	(3) Clearance
		(4) Usage of tool to assemble a single part
Semi-finished product	Supporting	(5) Obstacles generated when a part is assembled
		(6) Unsteadiness generated when a part is assembled
		(7) Obstacles generated when a support part is used
		(8) Unsteadiness generated when a support part is used
	Movement	(9) Rotate to assemble a part easily
Support parts	Positioning	(10) Direction or side to use a support part
		(11) Size and weight
	Method to use a support part	(12) Clearance
		(13) Usage of tool to use support parts

3. Construction of data matrix based on parts structure using design information by CAD

3.1. Matrices to express the characteristic relationship between parts

From the design information in CAD, matrices that express the connection between parts and the evaluation values of the factors of work difficulty related to the structure of the parts were constructed. Several types of matrices were created to store the structure of the parts from CAD. Subsequently, these matrices were used to determine the assembly order of the parts.

The following matrices were created to represent the characteristic relationship between parts:

(m1) A matrix to present the connection between parts with the direction to assemble the part,

(m2) A matrix to present the characteristics of obstacles that appear when a single part is assembled,

(m3) A matrix to present the relationship between parts causing an unstable state of the semi-finished product when a single part is assembled.

In matrix (m1), part numbers are assigned to both rows and columns, whereas $\pm x$, $\pm y$, and $\pm z$ are allocated in the elements of row i and column j to present the direction in which part p_i at row i connects to part p_j at column j. The directions of $\pm x$, $\pm y$, and $\pm z$ are denoted in Figure 1.
In matrix (m2), part numbers are assigned to both rows and columns, and the evaluation value of the connection between parts p_i and p_j is described in the element of row i and column j. When no contact exists between these parts, 0 is allocated to the element.
In matrix (m3), part numbers are assigned to both rows and columns. If part p_j of column j is not included in the semi-finished product and the semi-finished product is determined as an unstable state when part p_i is about to be assembled, then the evaluation values are assigned to the element of row i and column j. Otherwise, 0 is assigned to the element.

3.2. Construction of characteristic relationships between parts using CAD

3.2.1 Structure including no inclusive relationship between parts

We developed a process to create the matrices from the relationship between parts in three-dimensional (3D) CAD. The 3D-CAD software used was Rhinoceros6, and the software to create the matrices was generated using Python. In this CAD, the position of a shape's vertex for each part and the vertex positions of a cube created from the outermost circumference of the part can be obtained.
Figure 1 shows the schematic diagram of the structure of parts including no inclusive relationship between parts. In a cube created from the outlines of the parts, vertices having minimum values in the x, y, and z-coordinates are regarded as the origin, and the vertices are denoted as Point10 ($x10$, $y10$, $z10$) in Figure 1. The direction vectors $dx1$, $dy1$, and $dz1$ in the x, y, and z-coordinates were investigated by comparing the origin with the vertices in the x, y, and z-directions.
The x vertices and the direction vector in the x-coordinates were compared to evaluate the connection. Under the condition that $x10 \geq x20$, the connection between different parts in the x-direction vector was assessed based on the following conditions:

(x1) $x10 - x20 < dx2 - \delta$,

(x2) $dx2 - \delta \leq x10 - x20 \leq dx2 + \delta$,

(x3) $x10 - x20 > dx2 + \delta$.

Here, δ denotes a small number. Because the overlapping area of two different parts is determined from these conditions, either the connection, contact, or noncontact of these parts is evaluated in the x-direction. In addition, similar comparisons are made in the y and z-directions to determine the relationship between the parts. Finally, the connection, contact, or noncontact relationship between the two parts is determined by evaluating the relationships between the parts in the x, y, and z-coordinates simultaneously.
Furthermore, product structures that have internal relationships between parts exist. However, if the overlapping area of two parts is evaluated in the product with the internal relationship, the internal relationship in the parts cannot be identified, and the relationship is regarded as the connection relationship between parts. Hence, engineers are required to modify the relationship between the parts with the internal relationship from connection to noncontact, as shown in Figure 2.

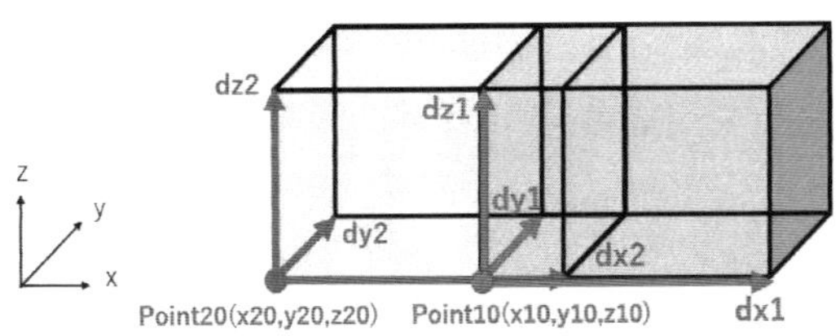

Figure 1. Schematic diagram of parts structure including no inclusive relationship of parts.

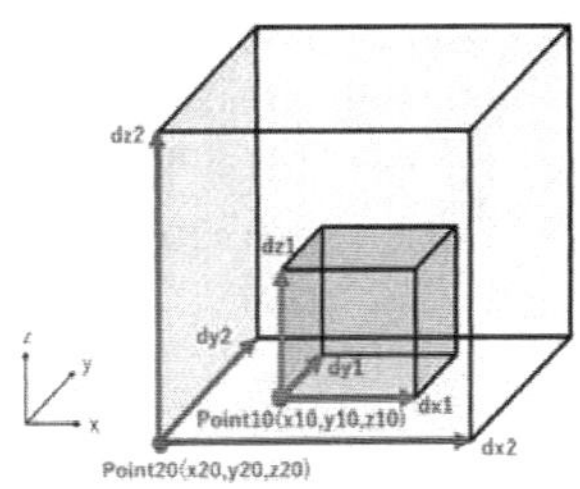

Figure 2. Schematic diagram of parts structure including inclusive relationship of parts.

3.3. Evaluation of factors of work difficulty in assembly process

3.3.1 Evaluation of obstacles in assembling a single part
We address the presence of obstacles, which is a factor causing work difficulty, when a single part is assembled. The difficulty caused by the contact relationship between the part that becomes an obstacle at the time of assembling a single part to the semi-finished product and the single part is classified into the following four levels:

L1: When no connection and no contact exist between the parts, the evaluation value is 0.
L2: When the parts are contacted at points or edges, the evaluation value is 10.
L3: When the parts are contacted at the surface of each of the other parts, the evaluation value is 50.
L4: When the parts are connected, the evaluation value is 0.

3.3.2 Evaluation of unstable state of semi-finished products at the time of assembling a single part
The unstable state of a semi-finished product at the time of assembling a single part is regarded as a factor of work difficulty. The unstable state is typically caused when the parts required to yield a stable state are not assembled to the semi-finished product. For example, an unstable state of a semi-finished product is generated by the presence of unconnected parts at a lower position than the assembling position of a part. The following evaluation values are used to evaluate the unstable state:

L1: When the semi-finished product does not include more than one part required to yield a stable state, the condition is judged as an unstable state and the evaluation value is 100.
L2: When the semi-finished product includes all parts required to yield a stable state, the condition is judged as a stable state and the evaluation value is 0.

3.3.3 Evaluation of assembling direction of parts
The following evaluation is performed from the assembling direction of a single part when the part is assembled to a semi-finished product. Here, the right hand is assumed as the dominant arm.

L1: When the single part is assembled in the +*x*-direction, the evaluation value is 100.
L2: When the single part is assembled in the -*y*-direction, the evaluation value takes 50.
L3: When the single part is assembled in the +*z*-direction, the evaluation value takes 10.
L4: When the single part is assembled in the -*x*, +*y*, and -z-directions, the evaluation value is 1.

All of the *x*, *y*, and *z*-directions were determined according to the coordinate system shown in Figure 1.

3.3.4 Evaluation of weight of parts to be assembled
At the time of assembling a single part to a semi-finished product, the weight of the part to be assembled is evaluated according to the following conditions:

L1: For 5.6 kg or more, the evaluation value is 1.5.
L2: For 3.4–5.6 kg, the evaluation value is 1.2.
L3: For 1.1–3.4 kg, the evaluation value is 1.1.
L4: For 1.1 kg or less, the evaluation value is 1.0.

4. Method for determining parts assembly order considering work difficulty

4.1. Construction of mathematical model
We developed a method to determine the assembly order of parts to facilitate the work. The work difficulty is evaluated from the factors when a single part is assembled to a semi-finished product in Table 1. The order of parts to assemble the product is generated by minimizing the sum of the values of the difficulty when all parts are assembled. Equation (1) shows the objective function in the mathematical model.

$$\min\ z = \sum_{j=1}^{J}\sum_{i=1}^{I} s_j D_{ij} \alpha_{ij} n_{ij} + \sum_{j=1}^{J}\sum_{i=1}^{I} s_j D_{ij} F_{ij} b_{ij} \qquad (1)$$

In Equation (1), *i* denotes the part number to determine the order of parts to assemble the product. n_{ij} is the number of obstacles when part *j* is assembled under the condition that part *i* is assembled prior to

part j. α_{ij} is the evaluation value of part i as an obstacle from the point of view of part j. From these symbols, the first term of Equation (1) denotes the sum of evaluation values related to obstacles in the assembly order of parts generated.
F_{ij} denotes the evaluation value related to the unstable state of the semi-finished product, in which part i is NOT assembled when part j is about to be assembled. b_{ij} is variable, which is 1 when part i is assembled prior to part j, and 0 otherwise. From these symbols, the second term of Equation (1) denotes the sum of evaluation values related to the unstable state of the semi-finished product. s_j denotes the evaluation values of the weight of part j. D_{ij} denotes the evaluation values of the direction of part j to assemble when part j is assembled to part i.
To generate the assembly order of parts, the evaluation values of obstacles in the assembly order of parts and the unstable state of the semi-finished product are simultaneously reduced by minimizing Equation (1). By minimizing the objective function, we assume that the assembly order of parts that removes the difficult state to work to the maximum is obtained.
The evaluation values of α_{ij}, F_{ij}, D_{ij}, and s_j are as shown in Sections 3.3.1, 3.3.2, 3.3.3, and 3.3.4, respectively. These variables are treated as the factors of difficulty.

4.2. Numerical experiment
A mathematical model was used to calculate the assembly order of parts. A product composed of LEGO blocks was used as the final product. Figure 3 shows a photograph of the product, and the number of parts of the product was 42.
Numerical experiments were performed for each of the semi-finished products A, B, C, and D shown in Figure 3. The weight and direction of the parts to be assembled were assumed to be the same for all parts because of the small weight of all blocks and the downward direction, and both evaluation values were set to 1. Table 2 shows the objective function values obtained from the numerical experiments.

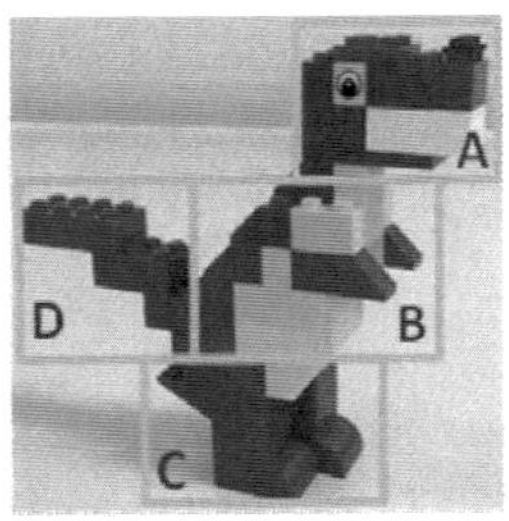

Figure 3. External form of the product used for assembly experiment.

Table 2. Value of objective function obtained using mathematical model.

Semi-finished product		Value of terms of objective function: Obstacles	Value of terms of objective function: Unstable state of the semi-finished product	Objective function
A	Min	9050	2500	11550
	Max	14650	3650	18300
B	Min	15750	5150	20900
	Max	36250	6800	43050
C	Min	12550	2700	15250
	Max	23850	4150	28000
D	Min	400	200	600
	Max	600	300	900
Assembly process of the final product	Min	1000	500	1500
	Max	1550	700	2250

5. Estimation of work time using MTM and evaluation of proposed system

The work time was estimated using the MTM method [4] from the assembly order of parts obtained in the numerical experiment. Because the parts assembly process is described as "repeating a single part assembled to a semi-finished product," the process for assembling a single part is regarded as the unit work process, and the operation time is calculated from motions included in the unit work process using MTM. The operation time is estimated according to the following procedure:

(1) The basic motion in the unit work process is composed of M (MOVE), G (GRASP), M, P (POSITION), P, and M.

(2) The MTM representation related to the distance from the work area to the parts storage space is generated as the same data in all unit works.

(3) The differences in work motion modified by the assembly order of parts from the basic work motion presented in (1) are evaluated, and MTM representation is modified.

(4) The operation times of all work units are calculated using the MTM representation modified in (2). Subsequently, the total operation time is estimated from all work units.

Table 3 shows the operation time calculated by MTM for the assembly processes of different semi-finished products. In this table, "max: and "min" denote the operation times of the parts assembly process obtained at the maximum and minimum of the objective function, respectively.

From the results of products A, B, and C, it was shown that the operation time calculated from MTM was not necessarily short even in the case of the work sequence at the minimum objective function.

We speculate that this result was caused by the following characteristics in the calculation of MTM: the unstable state of the semi-finished product was not considered, and the effects of obstacles were evaluated to be small owing to the simple structure of the product, as the evaluation of the unstable state of the semi-finished product was not included in MTM method originally.

Figure 4 shows the results of the experiment using the actual product based on the assembly order of the parts obtained from the numerical experiment. Six participants were recruited for the experiment. Three participants performed the assembly work thrice using the order of the parts obtained from the minimization of the objective function, followed by the assembly work thrice using the order obtained from the maximization of the objective function three times. These participants were named the A-Group. The remaining three participants performed the work with the opposite order related to the objective function and were named the B-Group.

Figure 4 shows that the operation time of the work obtained from the minimization of the objective function was shorter than that obtained from the maximization of the objective function, and that the learning effect of the work obtained from the minimization of the objective function was higher than those of other works. The results demonstrate the effectiveness of the proposed mathematical model for calculating the assembly order of parts.

The total operation time estimated by MTM differed by approximately 25 s from that of the assembly experiment. This difference was assumed to be due to the non-consideration of the unstable state in the calculation using MTM. However, the difference in the time required (less than 20% in 170 s in the experiment) was considered to be effective for practical use.

Table 3. Operation time estimated by MTM for different assembly processes on different products.

	Estimated time (s)					
Semi-finished product	A	B	C	D	Assembly process of the final product	Total
Min	32.414	48.919	36.865	8.511	9.344	136.053
Max	32.011	47.307	35.253	8.511	9.344	132.426

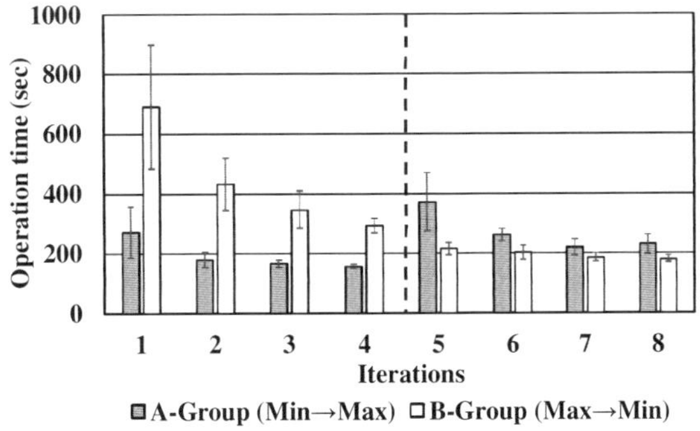

Figure 4. Results of experiment using the actual product based on assembly order of parts obtained from numerical experiments.

6. Summary

In this study, we developed a data structure to relate the characteristics of parts using CAD information and a method to generate the assembly order of parts considering work difficulty in the parts assembly process. In addition, we developed a system to estimate the operation time of the generated assembly order of the parts using MTM. These procedures were combined to form a total information system. The effectiveness of the total information system was demonstrated through numerical and assembly experiments.

In future work, we will investigate the effectiveness of the system for products with more complex component structures.

7. References

[1] Iwamoto H 2002 *Cell Production Systems* The Nikkan Kogyo Shinbun, Ltd. (in Japanese)

[2] Kadota T and Sakamoto S 1991 Methods Analysis and Design, *Handbook of Industrial Engineering (2nd Ed.)* Wiley Interscience pp 1415-1445

[3] Arakawa M 2009 Development of Information System for Product Design with Evaluating Assembly Process and Characteristics of Product Functions *Proceedings of Asia Pacific Industrial Engineering and Management Systems Conference 2009 (APIEMS2009)* Kitakyushu, Japan, pp 1694-1703

[4] Sellie C.N. 1991 Predetermined Motion–Time Systems and the Development and Use of Standard Data *Handbook of Industrial Engineering (2nd Ed.)* Wiley Interscience pp 1639-1698

Acknowledgments

This work was supported by Grant-in-Aid for Scientific Research (C) (18K04606) from the Japan Society for the Promotion of Science.

IOP Conf. Series: Materials Science and Engineering **947** (2020) 012003 doi:10.1088/1757-899X/947/1/012003

Ridge Penalization-based weighting approach for Eco-Efficiency assessment: The case in the food industry in the United States

Adeeb A Kutty, Galal M Abdella and Murat Kucukvar

Department of Mechanical and Industrial Engineering, Qatar University, Qatar

aa1704810@qu.edu.qa, gmg5005@qu.edu.qa , mkucukvar@qu.edu.qa

Abstract. Eco-efficiency assessment is of great importance for monitoring and managing environmental and economic aspects of sustainable development. The eco-efficiency indicators are required to assess and measure the impact of multiple environmental aspects per unit of economic value-added. The aggregation of multiple environmental impacts in the presence of high correlation is a critical challenge to sustainability practitioners. This study presents a weighting approach using ridge penalization-based regression to overcoming the consequence of the high correlation among the environmental aspects and hence providing accurate weighting values. The performance of the proposed approach is assessed using economic and environmental footprints of 20 food industries in the United States. The new weighting approach is expected to provide decision-makers with a quantitative management tool for monitoring and controlling core operational functions associated with the sustainable development and management.

1. Introduction

The eco-efficiency assessment is widely recognized as a powerful management tool for managing environmental sustainability aspects and enhancing the opportunities of the well-being of future generations [1]-[6]. The alignment with the sustainable development goals of the United Nations has recently become the focus of governmental and business organizations at both national and international levels [7]. The eco-efficiency assessment with high dimensional space of environmental impacts imposes a critical challenge to sustainability practitioners in specifying the weight of each environmental indicator to the eco-efficiency value [8],[9].

Several weighting techniques have been proposed and examined in literature, for instance, not limited to, linear programming [10],[11], Principal Component Analysis (PCA), Data Envelopment Analysis (DEA), and Factor Analysis (FA) [12]-[14], Regression Analysis [15]. The equal weighting (EW) is the most common among the existing methods [16]-[18]. Despite the distinctive advantage of mathematical and operational properties, this method has been extensively criticized for the lack of considering the double-counting when multiple indicators measure the same behavior [19].

The extension of statistical methods to the sustainability assessment context has received increasing attention over the recent years; see, for instance [20],[21]. The PCA, DEA, and FA are widely recognized for their ability to accommodate high dimensional space of sustainability indicators. Moreover, these methods are independent of subjective opinions [4]. The PCA is mainly based on the development of the Principal Components (PCs) as a linear combination of the corresponding sustainability indicators, then use their associated weight to complete the aggregation step in order to obtain a single value representing the overall environmental impact of these indicators [22].

The collinearity among two or more of the sustainability indicators describes the extent of linear correlation between the variables [23] thus, critical to the outcome of several of the existing weighting methods. The PCA is extensively used in literatures due to its capability in effectively handling the collinearity among the sustainability indicators [3],[24],[25]. Despite the merits, PCA lacks in interpreting the results of the dimension reduction analysis.

The PCs are linear combinations of all original indicators. A large number of independent variables can result in numerous significant coefficients in the first few PCs. The matter makes these PCs difficult to explicate [23],[26],[27]. Moreover, despite that the PCA is preferred for not relying on subjective and arbitrary opinions, it can be criticized for ignoring the relationship between the independent and dependent variables, especially as a weighting method. The inclusion of this relationship would provide a second criterion, in addition to the variation of the data matrix, to precisely quantify the individual weight of each of the sustainability indicators.

The PCA assigns a high amount of variance to the PC with the largest scale, the matter that results in undesirable skewness in the outcome. The normalization is a very well-known step to overcoming this issue. A difficulty that may result due to the normalization of the data matrix is that the number of PCs increases leading to difficulties in interpreting the results.

In accordance with the above, this paper presents a systematic methodology for eco-efficiency assessment using the ridge penalized regression to overcome the multicollinearity among the sustainability indicators. The ridge penalized regression is widely recognized in statistic for its effectiveness in overcoming the effect of multicollinearity on the accuracy and stability of the regression model. This study uses a dataset that represents the environmental impact of 20 food industries in the United States.

2. Methods

2.1. Input-Output (I-O) Model

The single region industry-by-industry I-O model is used here based on the Eora database that is connected with the UN's System of National Accounts and COMTRADE databases [21],[28]. In this study, the domestic supply and use tables (SUTs) of the U.S. economy were combined with several sustainability indicators. Then, the I-O model is used to quantify the economic (value-added), and environmental impacts of 15 food consumption industries in US; see Table 1.

Table 1. Food industrial sectors

No.	Industrial Category	Symbol
1	Beet sugar manufacturing	S1-BSM
2	Breakfast cereal manufacturing	S2-BCM
3	Cheese manufacturing	S3-CM
4	Coffee and tea manufacturing	S4-CTM
5	Dog and cat food manufacturing	S5-DCFM
6	Fats and oils	S6-FAO
7	Flour milling and malt	S7-FMM
8	Frozen food manufacturing	S8-FFM
9	Poultry processing	S9-PP
10	Seafood product	S10-SP
11	Snack food manufacturing	S11-SFM

Table 1. Food industrial sectors (Cont.)

No.	Industrial Category	Symbol
12	Soft drink and ice manufacturing	S12-SDIM
13	Sugar mills and refining	S13-SMR
14	Tortilla manufacturing	S14-TM
15	Wet corn milling	S15-WCM

2.2. Ridge Penalization-based Regression

The multiple regression analysis has been widely recognized in the literature as an effective tool to overcome the collinearity; see, for instance, [29]-[32]. The collinearity refers to the extent to which the indicators are linearly correlated to each other [33]. The multiple regression estimates the weights or relative-importance based on the extent to which each of the sustainability indicators significantly contributes to explaining the variability around the response variable. The penalization-based regression, in particular, has received notable attention as a weighting method; see [34],[35].

This paper uses a ridge penalization-based regression as a weighting method to overcome the multicollinearity phenomenon among the sustainability indicators. The error term ε, in the generalized linear relationship between response variable (y) and predictor variable (x) as shown in eq (1), is assumed to have a normal random distribution.

$$y_i = \beta_0 + \beta_1 x_{i1} + \beta_2 x_{i2} + \cdots + \beta_p x_{ip} + \varepsilon_i \qquad ; i = 1,2,\ldots \quad (1)$$

where $\boldsymbol{\beta_j}$ is the coefficient estimate associated with the jth indicator, and $\boldsymbol{p}$ is the number of indicators. The ridge-penalized regression is commonly formulated as a minimization problem of the squared errors were the problem is solved using the Ordinary Least Squared (OLS), Weighted Least Squared (WLS), or Maximum Likelihood Estimation (MLE) methods. The OLS and WLS are easier in practice than the MLE were the decision of selection depends on the practitioner. The ridge-based OLS formulation is as follows:

$$\widehat{\boldsymbol{\beta}}_{Ridge} = \underset{\beta_j \in \mathbb{R}^p}{\arg\max} \left\{ \sum_{i=1}^{n} \left(y_i - \beta_0 - \sum_{j=1}^{p} x_{ij}\beta_j \right) + \lambda \sum_{j=1}^{p} \beta_j^2 \right\} \quad (2)$$

where $\widehat{\boldsymbol{\beta}}$ is a vector that contains the estimated values of the regression coefficients, $\boldsymbol{n}$ is the number of observations, and λ is the tuning or shrinkage parameter, and its value is usually specified by K-fold cross-validation. Several computer packages available in the CRAN library and packages such as SPSS and Solver-Excel, can be used to solve the ridge-based OLS formulation, shown in (2). However, Figure 1 shows the outlines of the proposed methodology.

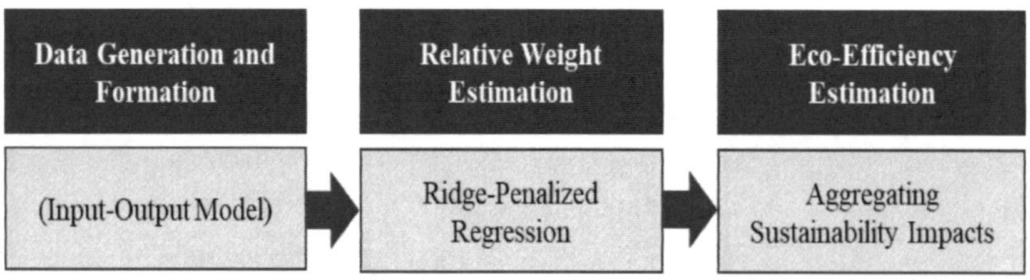

Figure 1. The proposed methodology.

3. Case study

3.1. Data generation and distribution

This study uses five sustainability indicators related to the food and beverage industry in U.S. These are: (1) CO_2 (Kt), (2) CO (Kt), (3) HFC-143a (Kt), (4) PM_{10} (Kt), (5) N_2O (Kt), and (6) SO_2 (Kt). The sustainability impacts of these indicators were estimated by using the Eora database-based economic input-output framework developed by [27] using the latest and high-resolution I-O tables of the U.S. economy; see also [36]. The household consumption (HC) under each food industrial category were calculated and used as the response variable. Figure 2 illustrates the distribution of the highest three impacts under each of the sustainability indicators.

3.2. Measuring Collinearity

This section is dedicated to measuring the level of collinearity among the sustainability indicators. The correlation of determination (R^2) is the most widely used method to measure the collinearity. Normalization step is neglected here due to the usage of same units for the selected indicators. In this study, R^2 explains the percentage of variation in one of the sustainability indicators that is predictable from the other indicators. The magnitude of the R^2 measure is limited between "0" and "1". The value "0" refers to a "*very-poor*" linear relationship, while the value "1" to a "*very-strong*" linear relationship.

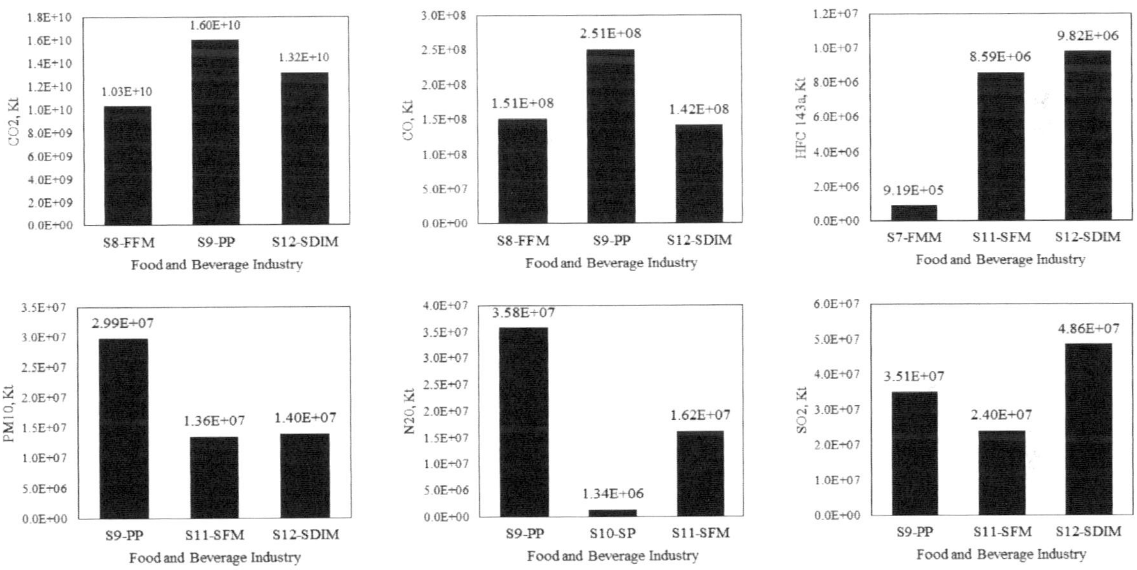

Figure 2. Distribution of the highest three-environmental sustainability impacts

The pairwise correlations among all the potential pairs of the sustainability indicators can be seen in Table 2.

Table 2. Correlation matrix for the selected sustainability indicators

	CO_2	CO	HFC	PM_{10}	N_2O	SO_2
CO_2	**1.000**	0.979	0.874	0.937	0.888	0.942
CO	0.979	**1.000**	0.830	0.975	0.957	0.866
HFC	0.874	0.830	**1.000**	0.829	0.754	0.922
PM_{10}	0.937	0.975	0.829	**1.000**	0.986	0.820
N_2O	0.888	0.957	0.754	0.986	**1.000**	0.726
SO_2	0.942	0.866	0.922	0.820	0.726	**1.000**

The correlation matrix shows a moderate to strong positive correlations ranging from 0.726 to 0.986. This finding would justify the need for ridge-penalization based regression.

3.3. Weighting sustainability indicators

To initiate the ridge-penalized regression, we used the Trace-Plotting method, proposed by [25], to determine the optimal value of λ. This method has been widely used under different research contexts; see, for instance, [22],[26]. Initially, the ridge regression coefficients are plotted over a wide range of λ. Secondly, we define the range of λ that exhibits better stability of the fitted regression coefficients. Finally, we select a single value of λ providing a better criterion of selection. However, The Mean Square Error (MSE), is used as a criterion for the selection of the optimal λ. Figure 3 shows the distribution of the ridge regression coefficients over a wide range of the tuning parameter λ. From Figure 3, one can easily notice the stability in the changes of the regression coefficients, namely the CO2, SO2, HFC-143a, and NO2, around the optimal λ. The best stability can be achieved when the λ value equals 0.014 to 1.4. The MSE at several values of λ has been estimated and the optimal value of λ is found to be 0.090 (MSE = 0.00092).

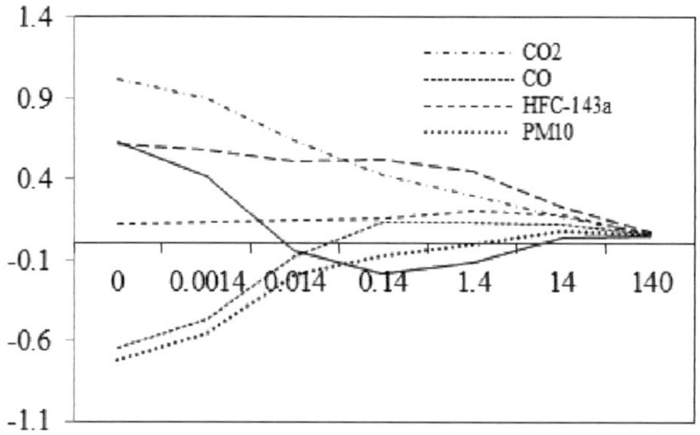

Figure 3. Changes of regression coefficients versus the tuning parameter

Table 3 shows the results of the analysis of variance (ANOVA) of the collected sample. In this study, the *p*-value (1.53E-08) is less than the set standard of 95% confidence interval, concluding the significance of the regression (R^2=0.995). While Table 4 summarizes the output of the ridge regression analysis. The results show that the SO2 has the largest positive impact on the regression model, while the PM10 has the smallest impact. The decision regarding the inclusion of the negative values in the aggregation step has been argued by several researchers [37].

Table 3. ANOVA Calculation

Source of Variation	d.f.	SS	MS	F	*p*-value
Regression	6	13.86	2.310	153.3	1.53E-08
Residual	9	0.135	0.015		
Total	15	14			

Table 4. Ridge Regression outcome

	CO_2	CO	HFC	PM_{10}	N_2O	SO_2
β_i	0.46	0.11	0.152	-0.081	-0.181	0.519
R^2	0.995					
Adjusted R^2	0.983					
Standard Error	0.122					
MSE	0.000092					

In this study, we replace the individual weight of the sustainability indicator by their associated relative weight (RW). The RW represents the importance of a specific indicator with regard to the other indicators. The RW is found as the absolute value of the individual weight divided by the sum of the absolute values of the individual weight of all the sustainability indicators. Table 5 reports the calculations of the weighting step.

Table 5. Weight Calculation

	CO_2	CO	HFC	PM_{10}	N_2O	NO_2
Absolute weight	0.46	0.11	0.152	0.081	0.181	0.519
Initial Rank	2	5	4	6	3	1
Relative Weight	0.31	0.07	0.10	0.05	0.12	0.35

3.4. Eco-Efficiency score calculation

The eco-efficiency is often calculated as the ratio between the economic value-added and the aggregation of the weighted impacts of the environmental indicators. Using the RW values reported in Table 4, we calculated the eco-efficiency scores for all the industrial categories and reported these scores in Figure 4.

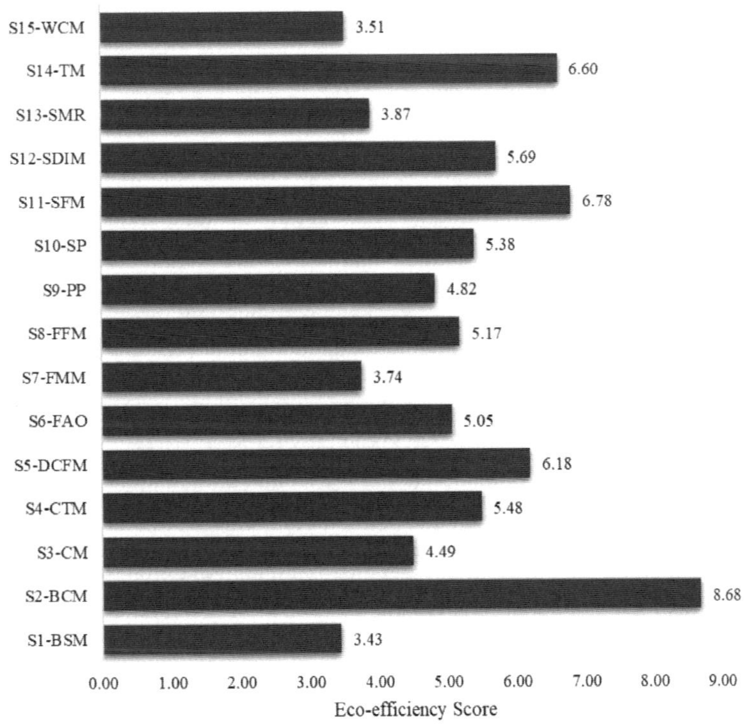

Figure 4. Eco-efficiency score distribution of the food and beverage industry

From Figure 4, S2-BCM has the highest eco-efficiency score, while the S1-BSM is the lowest. Several of the food and beverage industries have scored high scores, such as S14-TM, S11-SFM, and S5-DCFM. The eco-efficiency ratio or score, calculated can be referred to as the "*higher the better*" performance measure. This matter makes the comparison between the eco-efficiency performances of the industrial sectors difficult. However, in this paper, we use the average score of the eco-efficiency as a threshold between the "*Below-Average*," "*On-Average*" and "*Above-Average*" performance and specify the category of each industry based on its location with respect to the threshold value (5.26); see Table 6.

Table 6. Eco-Efficiency categories

Category	Above-Average	On-Average	Below-Average
Industry	S2-BCM S4-CTM S5-DCFM S10-SP S11-SFM S12-SDIM S14-TM	---	S1-BSM S3-CM S6-FAO S7-FMM S8-FFM S9-PP S13-SMR S15-WCM

The results in Table 5 show that 53.34% of the food industries are classified as "*Above-Average*" performance, while the rest are "*Below-Average*" performance. The results also show that none of the industrial categories is classified as "On-Average" performance.

4. Conclusion and Remarks

This research work introduced a penalization-based approach for estimating weights of sustainability indicators. Here, the importance of the penalization in reducing the impact of multicollinearity among the sustainability indicators during the aggregation step is emphasized. The results have shown that more than 50% of the food industries in the US are performing well in terms of eco-efficiency performance. The (BCM) has the best eco-efficiency performance, while the (BSM) has the worst performance comparing with the other indicators. However, in terms of the individual eco-efficiency performance, all the food and beverage industries have scored a value that is greater than 1.

For future research, variable selection methods, such as stepwise regression, can be used to identify the most significant indicators to be included in the weighting process [38]-[40]. The authors also suggest the extension of the adaptive LASSO-based thresholding to enhance the estimation of variance-covariance matrix of the PCA method. The new approach will be used later for developing of a composite indicator of eco-efficiency assessment, further details of the adaptive LASSO can be found in [41],[42]. For future research, the authors also suggest the use of hybrid life cycle sustainability assessment methods [40]–[55]; ecological footprint analysis [56]-[58]; and economic input-output analysis [59], combined with other decision making models such as fuzzy multi criteria decision making [60], forecasting [61], agent based modelling [62],[63], and system dynamic modelling [64]-[68] considering Triple Bottom Line (TBL) approach. Finally, the multivariate regression is another suggested approach [69]-[70] to complete the aggregation step and develop a single composite indicator for the sustainability assessment, ruling out the difficulty in finding an appropriate response variable.

References

[1] Brundtland Commission 1987 "Our Common Future"

[2] Garvare R and Isaksson R 2001 "Sustainable development: Extending the scope of business excellence models" *Meas. Bus. Excell.* **5** pp 11–15

[3] Park Y S, Egilmez G and Kucukvar M 2015 "A Novel Life Cycle-based Principal Component Analysis Framework for Eco-efficiency Analysis: Case of the United States Manufacturing and Transportation Nexus" *J. Clean. Prod.* **92** pp 327–42

[4] Onat N C, Kucukvar M and Afshar S 2019 "Eco-efficiency of electric vehicles in the United States: A life cycle assessment based principal component analysis" *J. Clean. Prod.* **212** pp 515–26

[5] Tatari O and Kucukvar M 2012 "Eco-Efficiency of Construction Materials: A Data Envelopment Analysis" *J. Constr. Eng. Manag.* **138** pp 733–41

[6] Egilmez G, Gumus G, Kucukvar M and Tatari O 2016 "A fuzzy data envelopment analysis framework for dealing with uncertainty impacts of input–output life cycle assessment models

on eco-efficiency assessment" *J. Clean. Prod.* **129** pp 622–36

[7] Onat N C, Kucukvar M, Halog A and Cloutier S 2017 "Systems Thinking for Life Cycle Sustainability Assessment: A Review of Recent Developments, Applications, and Future Perspectives," *Sustain.* **9**, p 706

[8] Abdella G M, Kucukvar M, Onat N C, Al-Yafay H M and Bulak M E 2020 "Sustainability assessment and modeling based on supervised machine learning techniques: The case for food consumption" *J. Clean. Prod.* **251**

[9] Park Y S, Egilmez G and Kucukvar M 2016 "Emergy and end-point impact assessment of agricultural and food production in the United States: A supply chain-linked Ecologically-based Life Cycle Assessment" *Ecol. Indic.* **62** pp 117–37

[10] Egilmez G, Kucukvar M, Tatari O and Bhutta M K S 2014 "Supply chain sustainability assessment of the U.S. food manufacturing sectors: A life cycle-based frontier approach," *Resour. Conserv. Recycl.* **82** pp 8–20

[11] Onat N C 2015 "Integrated sustainability assessment framework for the U.S. transportation," University of Central Florida

[12] Mouron P, Scholz R W, Nemecek T and Weber O 2006 "Life cycle management on Swiss fruit farms: Relating environmental and income indicators for apple-growing," *Ecol. Econ.* **58** pp 561–78

[13] Soler Rovira J and Soler Rovira P 2009 "Assessment of aggregated indicators of sustainability using PCA: the case of apple trade in Spain," in *6th ICLCA-AFS Towards a sustainable management of the food chain*

[14] Tatari O and Kucukvar M 2012 "Eco-Efficiency of Construction Materials: Data Envelopment Analysis," *J. Constr. Eng. Manag.* **138** pp 733–41

[15] Becker W, Saisana M, Paruolo P and Vandecasteele I 2017 "Weights and importance in composite indicators: Closing the gap," *Ecol. Indic.* **80** pp 12–22

[16] Egilmez G, Kucukvar M and Park Y S 2016 "Supply chain-linked sustainability assessment of the US manufacturing: An ecosystem perspective," *Sustain. Prod. Consum.* **5** pp 65-81

[17] Gumus S, Egilmez G, Kucukvar M and Park Y S 2016 "Integrating expert weighting and multi-criteria decision making into eco-efficiency analysis: the case of US manufacturing," *J. Oper. Res. Soc.* **67** pp 616–28

[18] Onat N C, Gumus S, Kucukvar M and Tatari O 2016 "Application of the TOPSIS and intuitionistic fuzzy set approaches for ranking the life cycle sustainability performance of alternative vehicle technologies," *Sustain. Prod. Consum.* **6**

[19] Reisi M, Aye L, Rajabifard A and Ngo T 2014 "Transport sustainability index: Melbourne case study," *Ecol. Indic.* **43** pp 288–96

[20] Yang L, Ouyang H, Fang K, Ye L and Zhang J 2015 "Evaluation of regional environmental efficiencies in China based on super-efficiency-DEA" *Ecol. Indic.* **51** pp 13–19

[21] Kucukvar M, Onat N C, Abdella G M and Tatari O 2019 "Assessing regional and global environmental footprints and value added of the largest food producers in the world," *Resour. Conserv. Recycl.* **144** pp 187–97

[22] Saisana M and Tarantola S 2002 "State-of-the-art Report on Current Methodologies and Practices for Composite Indicator Development,"

[23] Gan X et al. 2017 "When to use what: Methods for weighting and aggregating sustainability indicators" *Ecol. Indic.* **81** pp. 491–502

[24] Jiang Q et al. 2018 "A principal component analysis based three-dimensional sustainability assessment model to evaluate corporate sustainable performance," *J. Clean. Prod.* **187**, pp 625–37

[25] Iribarren D, Vázquez-Rowe I, Moreira M T and Feijoo G 2010 "Further potentials in the joint implementation of life cycle assessment and data envelopment analysis.," *Sci. Total Environ.* **408** pp 5265–72

[26] Jollife I T and Cadima J 2016 "Principal component analysis: A review and recent developments," *Phil. Trans. Roy. Soc. A: Math., Phy. Eng. Sci.* **374** no. 2065. Royal Society of London

[27] Sen B, Kucukvar M, Onat N C, and Tatari O 2019 "Life cycle sustainability assessment of autonomous heavy-duty trucks" *J. Ind. Ecol.*

[28] Onat N C, Kucukvar M, Halog A and Cloutier S 2017 "Systems thinking for life cycle sustainability assessment: A review of recent developments, applications, and future perspectives," *Sustain.* **9**

[29] Kim J, Abdella G M, Kim S, Al-Khalifa K N and Hamouda A M S 2019 "Control charts for variability monitoring in high-dimensional processes," *Comput. Ind. Eng.* **130** pp 309–16

[30] Abdella G M, Al-Khalifa K N, Tayseer M A and Hamouda A M S 2019 "Modelling trends in road crash frequency in Qatar State," *Int. J. Oper. Res.*

[31] Al Sheeb B, Abdella G M, Hamouda A M S and Abdulwahed M S 2019 "Predictive modeling of first-year student performance in engineering education using sequential penalization-based regression," *J. Stat. Manag. Syst.* **22** pp 31–50

[32] Abdella G M, Alhajyaseen W, Al-Khalifa K N and Hamouda A M S 2017 "Usage of nonlinear regression for modeling the behavior of motor vehicle crash fatality (MVF) rate". In: The Proceedings of the International Conference on Industrial Engineering and Operations Management, Rabat, Morocco, pp 1827–34

[33] Abdella G M, Kim J, Al-Khalifa K N and Hamouda A M S 2019 "Penalized Conway-Maxwell-Poisson regression for modelling dispersed discrete data: The case study of motor vehicle crash frequency," *Saf. Sci.* **120**, pp 157–63

[34] Azzurra A, Massimiliano A and Angela M 2019 "Measuring sustainable food consumption: A case study on organic food," *Sustain. Prod. Consum.* **17** pp 95–107

[35] Abdur Rouf K B, Abdella G M, Al-Khalifa K N and Alhajyaseen W 2018 "Ridge penalization-based generalized linear model (GzLM) for predicting risky-driving index". In the Proceedings of the International Conference on Industrial Engineering and Operations Management, Washington DC, USA pp. 1462–73

[36] Bennbea S, Wazwaz A, Abujarbou A, Abdella G M and Musharavati F 2018 "Towards Sustainable Society: Design of Food Waste Recycling Machine". In the Proceedings of the International Conference on Industrial Engineering and Operations Management, Bandung, Indonesia, March 6-8, pp. 2495–2508

[37] Saisana M and Philippas D 2012 Sustainable Society Index (SSI): Taking societies' pulse along social, environmental and economic issues.

[38] Abdella G M, Al-Khalifa K N, Kim S, Jeong M K, Elsayed E A and Hamouda A M S 2017 "Variable Selection-based Multivariate Cumulative Sum Control Chart," *Qual. Reliab. Eng. Int.* **33** pp. 565–78

[39] Wang K and Jiang W 2009 "High-dimensional process monitoring and fault isolation via variable selection," *J. Qual. Technol.* **41** pp 247–58

[40] Jiang W, Wang K and Tsung F 2012 "A variable-selection-based multivariate EWMA chart for process monitoring and diagnosis," *J. Qual. Technol.* **44** pp 209–30

[41] Abdella G M, Kim J, Kim S, Al-Khalifa K N, Jeong M K, Hamouda A M S and Elsayed E A 2019 "An adaptive thresholding-based process variability monitoring," *J. Qual. Technol.* **51** pp. 242–56

[42] Cai T and Liu W 2011 "Adaptive thresholding for sparse covariance matrix estimation," *J. Am. Stat. Assoc.* **106** pp 672–84

[43] Kucukvar M, Onat N C and Haider M A 2018 "Material dependence of national energy development plans: The case for Turkey and United Kingdom," *J. Clean. Prod.* **200** pp 490–500

[44] Sen B, Onat N C, Kucukvar M and Tatari O 2019 "Material footprint of electric vehicles: A

multiregional life cycle assessment," *J. Clean. Prod.* **209** pp 1033–43

[45] Onat N C, Kucukvar M, Aboushaqrah N N M and Jabbar R 2019 "How sustainable is electric mobility? A comprehensive sustainability assessment approach for the case of Qatar," *Appl. Energy.* **250** pp 461–77

[46] Cloutier S et al. 2017 "Measures of a sustainable commute as a predictor of happiness," *Sustain.* **9**

[47] Onat N C, Kucukvar M and Tatari O 2015 "Conventional, hybrid, plug-in hybrid or electric vehicles? State-based comparative carbon and energy footprint analysis in the United States," *Appl. Energy.* **150** pp 36–49

[48] Onat N C, Kucukvar M and Tatari O 2014, "Integrating triple bottom line input-output analysis into life cycle sustainability assessment framework: The case for US buildings," *Int. J. Life Cycle Assess.* **19** pp 1488–1505

[49] Onat N C, Kucukvar M and O Tatari 2014 "Towards life cycle sustainability assessment of alternative passenger vehicles," *Sustain.* **6**

[50] Tatari O and Kucukvar M 2012 "Sustainability assessment of US construction sectors: ecosystems perspective" *J. Const. Eng. Manag.* **138**. pp 918-922.

[51] Kucukvar M and Tatari O 2012 "Ecologically based hybrid life cycle analysis of continuously reinforced concrete and hot-mix asphalt pavements" *Trans. Rese. Part D: Tran. Env.* **17**. pp 86-90

[52] Kucukvar M, Egilmez G and Tatari O 2016 "Life cycle assessment and optimization-based decision analysis of construction waste recycling for a LEED-certified university building" *Sust.* **8**. p 89

[53] Kucukvar M and Samadi H 2015 "Linking national food production to global supply chain impacts for the energy-climate challenge: the cases of the EU-27 and Turkey" *J. Clean. Prod.* **108**. pp 395-408

[54] Park Y S, Egilmez G and Kucukvar M 2017 "Cradle-to-gate Life Cycle Analysis of Agricultural and Food Production in the US: A TRACI Impact Assessment" **274**. *Sust. Agrofood Sec.*

[55] Kucukvar M, Cansev B, Egilmez G, Onat N C and Samadi H 2016 "Energy-climate-manufacturing nexus: New insights from the regional and global supply chains of manufacturing industries" *Appl. Energy* **184**. pp 889-904

[56] Kucukvar M, Onat N C, Abdella G M and Tatari O 2019 "Assessing regional and global environmental footprints and value added of the largest food producers in the world" *Res. Conserv. Recycl.* **144**. pp 187-197

[57] Kucukvar M, Ismaen R, Onat N C, Al-Hajri A, Al-Yafay H and Al-Darwish A 2019 "Exploring the social, economic and environmental footprint of food consumption: a supply chain-linked sustainability assessment" *In 2019 IEEE 6th International Conference on Industrial Engineering and Applications (ICIEA)* pp 733-742

[58] Kucukvar M, Haider M A and Onat N C 2017 "Exploring the material footprints of national electricity production scenarios until 2050: the case for Turkey and UK" *Res. Conserv. Recycl.* **125.** pp 251-263

[59] Egilmez G, Kucukvar M and Tatari O 2013 "Sustainability assessment of US manufacturing sectors: an economic input output-based frontier approach" *J. Clean. Prod.* **53**. pp 91-102

[60] Onat N C, Kucukvar M, O Tatari and Zheng Q P 2016 "Combined application of multi-criteria optimization and life-cycle sustainability assessment for optimal distribution of alternative passenger cars in U.S.," *J. Clean. Prod.* **112** pp 291–307

[61] Shaikh M A, Kucukvar M, Onat N C and Kirkil G 2017 "A framework for water and carbon footprint analysis of national electricity production scenarios," *Energy.* **139**

[62] Onat N C, Noori M, Kucukvar M, Zhao Y, Tatari O and Chester M 2017 "Exploring the suitability of electric vehicles in the United States," *Energy.* **121**

IOP Conf. Series: Materials Science and Engineering **947** (2020) 012003 doi:10.1088/1757-899X/947/1/012003

[63] Noori M, Zhao Y, Onat N C, Gardner S and Tatari O 2016 "Light-duty electric vehicles to improve the integrity of the electricity grid through Vehicle-to-Grid technology: Analysis of regional net revenue and emissions savings," *Appl. Energy.* **168** pp 146–58

[64] Alirezaei M, Onat N C, Tatari O and Abdel-Aty M 2017 "The Climate Change-Road Safety-Economy Nexus: A System Dynamics Approach to Understanding Complex Interdependencies," *Systems.* **5** p 6

[65] Kelly C, Onat N C and Tatari O 2019 "Water and carbon footprint reduction potential of renewable energy in the United States: A policy analysis using system dynamics," *J. Clean. Prod.* **228** pp 910–26

[66] Ercan T, Onat N C and Tatari O 2016 "Investigating carbon footprint reduction potential of public transportation in United States: A system dynamics approach," *J. Clean. Prod.* **133**

[67] Onat N C, Kucukvar M and Tatari O 2016 "Uncertainty-embedded dynamic life cycle sustainability assessment framework: An ex-ante perspective on the impacts of alternative vehicle options," *Energy.* **112** pp 715–28

[68] Onat N C, Kucukvar M, Tatari O and Egilmez G 2016 "Integration of system dynamics approach toward deepening and broadening the life cycle sustainability assessment framework: a case for electric vehicles," *Int. J. Life Cycle Assess.* **21**

[69] Abdella G M, Yang K and Alaeddinin A 2014 "Multivariate adaptive approach for monitoring simple linear profiles," *Int. J. Data Analysis Techniques and Strategies* **6**

[70] Abdella G M, Kim J, Al-Khalifa K N and Hamouda A M S 2016 "Double EWMA-based polynomial profile monitoring" *Int. J. Quality and Reliability* **32** pp 2639–52

IOP Conf. Series: Materials Science and Engineering **947** (2020) 012004 doi:10.1088/1757-899X/947/1/012004

Machinery maintenance model for evaluating and increasing maintenance, repairs and operations within Industry 4.0 concept

P Poór[1], J Basl[2]

[1,2]Faculty of Mechanical Engineering, Department of Industrial Engineering and Management, University of West Bohemia, Plzeň, Univerzitní 8, 301 00, Czech Republic

[1]poorpeter@gmail.com, [2]basljo@kpv.zcu.cz

Abstract. Machinery maintenance is a crucial part of every manufacturing organization. As production, as maintenance evolved through the history. The upcoming 4th Industrial revolution has also impact on machinery and equipment in companies. Companies have already started adopting new approaches so machinery maintenance, based on various levels of readiness and maturity. This paper focuses primarily on creating a model for Maintenance 4.0. The aim is to determine key "components" or defining factors of new approaches to machinery maintenance in accordance with principles of Industry 4.0. implementation.

1. Introduction

The term Industry 4.0 conceals the current trend of digitization and production automation. Great social, political and economic changes in our memories are often connected to term "revolutionary". Not only led by politicians or generals, also by physicists and other science figures, technicians and engineers, inventors or entrepreneurs.

In 1837 the French revolutionary and socialist theorist L. A. Blanqui probably used the term "industrial revolution" as the first to compare the economic and social development [1]. The first industrial revolution happened at the end of the 18th century and J. Watts steam engine became a symbol of it. The second industrial revolution is driven by electrification and motorization, started less than a hundred years later in 1870. Electric power, mainly associated with the names of T.A. Edison, N. Tesla, W. Siemens and internal combustion engines (N. Otto, GW Daimler, K. Benz, R. Diesel, etc.) became the symbols of this period. Key points of the third industrial revolution are digitization and robotization. It is considered to begun in 1969 and is characterized by the advent of microprocessors and the use of computers.

The beginning of fourth industrial revolution dates back to 2011; but the main concepts were presented to the professional public in 2013 at the Hannover fair. It is represented by information technology and intelligent cyber-physical systems leading to the automation of production processes, gradually replacing the human workforce and provoking the demand for the establishment of modern workplaces. This will be helped by cameras, transmitters, sensors, code readers and other high-end technology devices that take over not only stereotyped and simple tasks and dangerous activities, but also intelligent decision-making processes that are currently being carried out by people. According to experts, this should lead to a productivity increase of up to one third compared to the current situation.

IOP Conf. Series: Materials Science and Engineering **947** (2020) 012004 doi:10.1088/1757-899X/947/1/012004

2. Maintenance as the tenth pillar of Industry 4.0.

Current academics [20,21,22] often talk about nine pillars of Industry 4.0., which "help" factories to be intelligent and that is what Industry 4.0 is all about.

- Horizontal and Vertical Integration

Horizontal integration connects cyber-physical systems and enterprise systems (BIS), where all devices in the facility are connected with each other. Vertical integration makes involves systems and humans, when interfaces (humans and machines communicate in two different ways) are used during vertical integration.

- Industrial Internet of Things

According to the world research company Gartner, the Internet of Things (IoT) can be defined as "a network of physical objects that contains built-in technologies for communication and perception; or affects their internal conditions or external environment" [2]. Accenture and the Bankinter Foundation of Innovation define IoT: "The Internet of Things consists of things connected to the Internet anytime, anywhere. In technical meaning the Internet of Things incorporates sensors and devices into the common objects that are connected to the Internet via wired or wireless networks. [3]. For the purpose of our research the concept of the Internet of Things describes that things can be connected to the Internet via transmission networks. Manufacturers are attaching sensors to machines and other physical assets on the plant floor to collect data which influences decisions real time and leads to increased efficiency and productivity.

- Big Data and Data analytics

Big Data and Data analytics is mostly used in industries to reduce downtime and wastages. Also, data evaluation helps reducing downtimes and increasing productivity. This is useful when companies are adopting principles of predictive maintenance.

- Simulation

A simulation is used to simulate a virtual environment of the factory (digital twins). The digital twin is a virtual copy of individual machines or entire, even very extensive production or logistics processes. It allows to model their activity and create variants as if it were a real machine or process.

- Additive Manufacturing and 3D Printing

Additive manufacturing (AM) is defined as the process of joining materials to produce objects from 3D data models, mostly layer by layer, as opposed to subtractive manufacturing methods [7]. The layering of the material is what fundamentally distinguishes additive production from other traditional production techniques. The material is layered using a laser or a print head [23]. Additive manufacturing would not be possible without the use of other technologies - CAD, CAM and CNC, thanks to which it is possible to create and print three-dimensional models and objects.

- Autonomous Robots

Industrial robots are generally designed to work for people and they perform activities in a restricted area according to relatively complex programs. A collaborative robot ('cobot') is robot an industrial robot that is designed to cooperate with human operators and create values together. A traditional robot is limited by safety cages and sensors that are often expensive and prevent easy access, but collaborative robots are accessible for human operators (easier maintenance [4], lower costs for safety). They are equipped with features and detection sensors that allow them to recognize where the human operators are. They can be easily programmed. While traditional industrial robots require advanced knowledge and programming experience, collaborative robots are able to teach themselves [4].

- Cloud Services

Cloud computing is a service implemented using the Internet (public IP network), represented by an imaginary network cloud, using virtual servers [5]. As a characteristic of this type of technology are speed of access, sharing of resources and high elasticity. Cloud computing is available in main three models:

SaaS - Software as a Service - allows the user to run software applications over the Internet without the need to own them (Gmail, Microsoft Online).

PaaS - Platform as a Service - provides the user with a computing platform for supporting web applications via the Internet (Google Apps).
IaaS - Infrastructure as a Service - enables the use of computer hardware and software, including operating and communication systems. The service provider takes care of installation until maintenance. The benefits of the Cloud include: remote data storage, hardware performance, great variety of options, cost, location independence, reliability of the Cloud network and maintenance, which together with the management is in the hands of the service provider [6].

- Cyber Security in Industry 4.0

Cyber security's main role in accordance with Industry 4.0. principles is to prevent hacking company's server and data and safeguarding the data and performance of the server. Any security breach could damage multiple areas of the business, from supply chain to operations.

- Augmented Reality

Augmented Reality uses real environment and masks it with digital information. It displays digital content in real word through a device, such as a mobile phone or special eyeglasses. By connecting and writing digital information directly to real objects, people can process both physical and digital signals.

These 9 aspects can be defined as core key elements of Industry 4.0. We added an another "tenth pillar" focused on machinery maintenance - Maintenance 4.0. Some recent research shown in the next section tries to present definitions of Maintenance 4.0.

3. Literature review on Maintenance 4.0.

Cachada et al. [8] consider maintenance as an integral part of the manufacturing process that contributes to the product quality, plant availability and ability to meet delivery schedules. This considers advanced and online analysis of collected data for earlier detection of the occurrence of possible machine failures, and supports technicians by providing a guided intelligent decision support. Galar & Kans [9] try to define maintenance with focus on Industry 4.0 as a smart manufacturing system consisting of physical assets and humans. Maintenance 4.0 is represented by self-learning and smart machines that predicts failure, makes diagnosis and triggers maintenance actions. The smart equipment is in form of embedded or cyber-physical systems (CPS), where physical and software components are intertwined. According to Algabroun et al. [10] Maintenance 4.0 should be able to execute data gathering, monitoring, analyzing, planning, and data presenting. These tasks should be performed in environment of vertical and horizontal integrated systems with end-to-end engineering solutions. To perform these tasks, Maintenance 4.0 must take into account necessary elements such as sensors, actuators, network, processors, middleware, databases, software and applications. Kandemir and Celik [11,12] view Maintenance 4.0 as one of the most futuristic subjects. By proper implementation higher availability performance, reduced downtimes, optimized energy consumption and savings on general maintenance costs can be achieved. They put big emphasis on people here. Skilled workers' adaptation into different roles and tasks must be provided. Furthermore, enhanced assessment skills for IT systems must be noted. Concept of Maintenance 4.0 has an important role to contribute future's maintenance standard and it lies in the core of industry 4.0. According to Jantunen et al. maintenance 4.0", comprises all topics related to the application of Industry 4.0's techniques and technologies to maintenance [13]. Chiu et al. [14] deal with Maintenance 4.0. only from point of view of Predictive maintenance (PdM). It monitors equipment health and indicates when a maintenance event will be required in the future, when it has been raised to the top priority as the key enabler for maximizing tool availability. Franciosi et al. [15] define Maintenance 4.0. with another crucial actor – sustainability. The role of Maintenance 4.0 for sustainable manufacturing is still in its infancy and it has big potential. However, there is no contrast between old-new or conventional-sustainable, but probably a potential synergy. Thus it is necessary to understand how maintenance engineering can be combined with big data potentiality in order to obtain the better "configuration" of Big Data Analytics.

4. Maintenance 4.0 model

Based on above-mentioned, we try to propose a Maintenance 4.0. model. Our model was initially spit into 3 main key concepts: **maintenance aspect** (Maintenance 4.0. "takes" features from previous forms of maintenance), **technology aspect** (describing advanced technologies to perform the maintenance itself) and **"people" aspect** (from TPM people are one of the crucial elements of maintenance). Before moving to describing the model itself, we realized that there is a need for **"other" aspect** of Maintenance 4.0. This was covered mostly by ecological and environmental aspects on company's environment.

The key idea of **Preventive Maintenance** is that "prevention is better than cure". An aid of intelligent software to schedule maintenance plan is also important. This is in a contrast with "old" maintenance approaches, where the main "strategy" was "run the device until its failure". **Predictive maintenance** can be expressed in many ways. For our model, RUL is the key factor. **Remaining useful life (RUL)** is defined as duration left for a system before it fails. Other successful models of determining predictive maintenance based on recent research are Physic, Data-driven, Probabilistic, Stochastic and A.I.-method based. Here, we are dealing with a certain time frame (weeks usually). Another part of Maintenance aspect is called **Prescriptive maintenance** and analytics. This is nowadays the most advanced option in maintenance. Based on big data, graph analysis, simulations, complex event processing, neural networks, heuristic and machine learning. This not only reflects possible results, but also evaluates which of approaches is the fastest or most efficient. With daily use of smartphones, tablets or other devices such as special data goggles mobile maintenance is emerging. Operators and technicians can track the status of a chosen device (group of devices, plant) through special mobile applications, which are part of complex CMMS, what brings us to "technological aspect" of Maintenance 4.0. model. Prognostic maintenance helps to predict time, when system or its part is no longer capable to maintain its function. This is built in abovementioned predictive analysis and maintenance, BUT unlike them, use of modern technologies (machine learning, pattern recognition, neural networks) is implemented here.

Internet of Things is a key part here. Devices and objects through the company ale linked and communicate together. From the technological point, 5G is crucial here (enables high data transfer at high speed and volumes), also combination of predictive maintenance with **Big Data** analysis. Big Data analysis provides a link information, which streams data from and to the company. Dealing with Maintenance 4.0. the hardest part will be how to filter data. Machinery health based on **condition monitoring** is here another key factor, where intelligent software plays a role. **EDGE computing** is a method to optimize cloud computing by performing data processing near the source of the data. Faster speed, safety and reducing of costs are the results of dividing between local and cloud data. **Collaborative robots**, which coexist and learn from humans with inspection drones and cleaning (maintenance) robots represent the **Artificial Intelligence** part of Maintenance 4.0. concept. Remotely controlled **drones** can repair damaged parts dangerous to humans. **Blockchain technology** enables higher level of automation, speeding up processes, cutting costs and decreasing interparty friction. **Machine learning** uses algorithms to detect anomalous patterns and identify threats of failure. With abovementioned CMMS and interconnected system **reporting of failures** is an automated process now.

"People" are more and more important with evolved forms of maintenance. This became evident since TPM implementation when operators are taught to care about machines as "their own". **Operators** played only a small or no-role in this process, but with Industry 4.0. technologies they now have to be fully involved in CMMS. But their role is not as important as role of **data scientists**. With technologies like Big Data they analyse complex patterns and provide advices based on them. People dealing with modern forms of maintenance need to be properly **trained**. Modern form of maintenance needs data scientists rather than "repairmen with a hammer". **collaboration** between all these processes is crucial.

We labelled the last part of model as "other". **Environmental protection** is a big thing now. Modern forms of maintenance try to achieve more efficient use of industrial plants and reduction in energy consumption and raw material wastage. Big data and predictive maintenance are catalysts for improving food efficiencies and **food waste and hunger reduction** is a result. By implementing

Maintenance 4.0. technologies we can achieve a significant reduction [19] of company's carbon footprint.

Table 1 Maintenance 4.0. model

Maintenance aspect		Technology		People		Other	
Preventive Maintenance	prevent failures before occur	Internet of things	Devices and objects linked together	Operators	fully involved in CMMS	Environment	efficient use of industrial plants and a reduction in energy consumption
Predictive maintenance	RUL,	5G	Very high data transfer	Data analysts	analyze complex patterns and provide advices		
Prescriptive maintenance and analytics	reflects possible results, also evaluates most efficient approaches	Big Data analysis	a link information, which streams data from and to the company	Specialists	"digitally assisted" partnership between man and machine	Food waste and hunger reduction	big data and predictive maintenance as catalysts for improving food efficiencies
		condition monitoring	Based on intelligent software				
Mobile Maintenance	use of smartphones, tablets or other devices data goggles.	EDGE computing	performing data processing near the source of the data	Training and onboarding	employees are considered as part of process		
		Drones assisted repairs	detect anomalous patterns and identify threats	Collaboration	employees' views are considered as part of the process	Carbon footprint	reduction by implementing M 4.0. technologies
		Robots	Collaborative, maintenance (cleaning)				
		Blockchain technology	automation, faster processes decreasing interparty friction	Remote monitoring (outsourcing)	leveraging the knowledge and expertise of trained experts		
		Artificial Intelligence	identify evolving failure and scheduling repairs				

5. Case study on preparedness of companies for a new type of maintenance processes

According to Haarman et al. [16] the most "critical" role of well-implemented maintenance in company is to predict failures. Only 11% are prepared for most advanced maintenance. Some companies have plans for this, or intentions in near future. But the truth is, that most companies have no plans due to their budget (main) or other reasons. For implementing most advanced forms of maintenance companies use data sources, sensors, environmental data, production controls or maintenance information systems. People are also an important part, they employ data scientists, IT specialists and reliability engineers. Almost 95% companies that implemented these stated that Maintenance 4.0. is responsible for improving in one or most cases. Main reasons why companies want to adopt Maintenance 4.0 are related to improve asset uptime (51%), which can be particularly important in circumstances like current economic upturn. Other important objectives include reducing costs (11%), risks associated with safety, health, environment and quality (SHEQ) (8%), extending asset lifetime (7%) and improving customer satisfaction (12%). Energy savings are barely mentioned, but "going green" will be very important.

From the point of view of data collection, most important were data types traditionally used in maintenance (usage of the asset' (from 72 to 83%) and maintenance history' (from 73 to 84%)), then data types mostly associated with more mature levels of predictive maintenance, which included condition of asset, condition data and history of assets from other company, and environmental data. Most data is collected from sensors, (77%), instrument inspections (75%) and interfaces with production control systems [17, 18]. From HW and SW point of view most common are: MS Excel,

Access, Database, Condition monitoring, Statistical software, WIFI, IoT, Mobile networks, Data-warehouse, Cloud.

Functions involved with Maintenance 4.0. from "traditional" technicians and maintenance inspectors to "modern" reliability engineers, data scientists, quality inspectors, IT specialists and others.

6. Conclusions

The ongoing 4th Industrial revolution resulted in changes on machinery and equipment maintenance. Modern technologies and approaches are presented in the article, where main aim was to propose a maintenance model incorporating these technologies and approaches with regard to Industry 4.0. principles. Maintenance has changed from "not let the device work until it goes wrong" to predictions (and taken measurements) of equipment degradation. The article presented Maintenance 4.0. as "tenth pillar" of Industry 4.0. Based on modern research review 3 main key aspects (maintenance, technology, "people") of model were introduced with 4th "other" aspect dealing mainly with environment and carbon footprint. From the case study it is obvious that companies that already implement most advanced forms of maintenance saw improvements in reducing costs, risks and improving processes and extending lifetime of their assets. However, this model is due to change, mainly because technical and maintenance aspects are still evolving and improving.

Acknowledgments

This research was supported by the project SGS-2018-031 Optimizing the parameters of a sustainable production system

7. References

[1] Coleman D C 1992 *Myth, history and the industrial revolution*. A&C Black.

[2] Gartner Inc 2017 IT Glossary - Internet of Things, http://www.gartner.com/it-glossary/internetof-things/ 2017

[3] Feller G 2011 The internet of things: In a connected world of smart objects. *Accenture & Bankinter* Foundation of Innovation, 24-29

[4] Poór P, Broum T, & Basl J 2019 Role of collaborative robots in Industry 4.0 with target on education in Industrial Engineering. In *2019 4th International Conference on Control, Robotics and Cybernetics (CRC)* (pp. 42-46). IEEE.

[5] Rittinghouse J W & Ransome J F 2016 *Cloud computing: implementation, management, and security*. CRC press.

[6] Alali F A & Yeh C L 2012. Cloud computing: Overview and risk analysis. *Journal of Information Systems,* **26(2)**, 13-33.

[7] Standard A S T M 2012 Standard terminology for additive manufacturing technologies. *ASTM International F2792-12a.*

[8] Cachada A, Barbosa J, Leitño P, Gcraldcs C A, Deusdado L, Costa J & Romero L 2018. Maintenance 4.0: Intelligent and predictive maintenance system architecture. In *2018 IEEE 23rd International Conference on Emerging Technologies and Factory Automation (ETFA)* (Vol. 1, pp. 139-146). IEEE.

[9] Galar D & Kans M 2017 The impact of maintenance 4.0 and big data analytics within strategic asset management. In *Maintenance Performance and Measurement and Management 2016 (MPMM 2016). November 28, Luleå, Sweden (pp. 96-104).* Luleå tekniska universitet.

[10] Algabroun H, Iftikhar M U, Al-Najjar B & Weyns D 2017 Maintenance 4.0 Framework using self-adaptive software architecture. In *Proceedings of 2nd International Conference on Maintenance Engineering, IncoME-II (pp. 299-309).*

[11] Kandemir C & Celik M 2017. Identifying training requirements to enhance basic skills for maintenance 4.0 in marine engineering through engine room simulator. In *13th International Conference on Engine Room Simulators (pp. 20-21).*

[12] Kandemir C & Celik M 2017. Determining shipboard integration requirements of maintenance 4.0 concept in marine engineering. *PressAcademia Procedia* **5(1)** 34-40.
[13] Jantunen E, Di Orio G, Larrinaga F, Becker M, Albano M, & Maló P. 2018. A framework for maintenance 4.0. In 2018 10th IMA International Conference on Modelling in *Industrial Maintenance and Reliability (MIMAR).* Institute of Mathematics and its Applications.
[14] Chiu Y C, Cheng F T, & Huang H C 2017. Developing a factory-wide intelligent predictive maintenance system based on Industry 4.0. *Journal of the Chinese Institute of Engineers,* **40(7)**, 562-571.
[15] Franciosi C, Iung B, Miranda S, & Riemma S 2018. Maintenance for sustainability in the industry 4.0 context: A scoping literature review. *IFAC-PapersOnLine,* **51(11)**, 903-908.
[16] Haarman M, de Klerk P, Decaigny P, Mulders M, Vassiliadis C, Sijtsema H, Gallo I.: Predictive maintenance 4.0 - beyond the hype: PdM 4.0 delivers results. *PricewaterhouseCoopers and Mannovation* 2018
[17] Pollák M, & Tkáč J 2019. Enterprise Information Data Management System for Small Manufacturing Company.
[18] Pollák M, Kaščak J, Telišková M, & Tkáč J 2019. Design of the 3D Printhead with Extruder for the Implementation of 3D Printing from Plastic and Recycling by Industrial Robot. *TEM Journal,* **8(3)**, 709.
[19] Orosz T, Poór P, Karban P and Pánek D "Power Transformer Design Optimization for Carbon Footprint," *2019 Electric Power Quality and Supply Reliability Conference (PQ) & 2019 Symposium on Electrical Engineering and Mechatronics (SEEM), Kärdla, Estonia,* 2019, pp. 1-4, doi: 10.1109/PQ.2019.8818261.
[20] Erboz G 2017 How to define industry 4.0: main pillars of industry 4.0. Szent Istvan University, Gödöllő, 1-9.
[21] Meet the nine pillars of Industry 4.0 and their relevance to industrial activity. 2020. from https://robotics.sumigusa.com/blogs/news/meet-the-nine-pillars-of-industry-4-0-and-their-relevance-to-industrial-activity
[22] Nine Pillars of IR 4.0 - Industry Revolution. 2020. From https://4irc.mediu.edu.my/industry-4-0/impact-4-0-industrial-revolution-in-education/

Methodology for evaluating the readiness of internal logistics processes for Industry 4.0

M Zoubek, M Šimon

Faculty of Mechanical Engineering, Department of Industrial Engineering and Management, University of West Bohemia, Pilsen, Univerzitní 8, 301 00, Czech Republic

zoubekm@kpv.zcu.cz, simon@kpv.zcu.cz

Abstract. The field of internal logistics is one of the critical areas of a company and has a great potential within the concept of Industry 4.0. The basic premise of smart production is to have a complete overview of all the logistics movements of products, i.e. from entry, through production to dispatch. Furthermore, this is exactly what the digitization of internal logistics allows. The development of Industry 4.0 based on digitization and automation presents excellent challenges for logistics and opportunities for further growth and development. Companies should optimize, automate and digitally transform internal logistics into a more autonomous and flexible form. However, a company must be prepared for this Industry 4.0 concept and know its level of maturity in this area. The primary purpose of this article is to describe a methodology for assessing the readiness of internal logistics processes, which are structured into several areas to cover the entire area of internal logistics. These processes are evaluated using the methodology, and the company's current level is assessed according to specific indicators for each area of internal logistics. By implementing elements from modern logistics, the aim is to increase the level of the company.

1. Introduction

Based on advanced digitization in industrial enterprises, it seems that the combination of internet technologies and technologies oriented in the field of intelligent objects (machines, production equipment, supporting equipment and products) is leading to a new fundamental paradigm shift in industrial production [1]. We are at the beginning of the Fourth Industrial Revolution (or Industry 4.0), which is already fundamentally influencing the way goods are produced and distributed. At the core of this revolution is the connection of the virtual cyber world with the world of physical reality [2]. Industry 4.0 is a digitization trend that focuses on automating production and creating so-called 'smart factories'. In these smart factories, intelligent machines, storage systems and production equipment in the form of cyber-physical systems are integrated into the production environment. These systems are able to independently exchange information and communicate, trigger actions, assess risks, control each other independently and make decisions [3]. The concept of Industry 4.0 also brings with it principles and essential technologies for autonomous functionality, which are applied to the production environment within smart factories. The main components used by Industry 4.0 include the Cyber-Physical System (CPS), the Internet of Things (IoT), Big Data and Data Mining (DM) and the Internet of Services (IoS) [4]. Cloud storage, augmented reality, autonomous robots, sensors and more are other key technologies and principles [5].

In addition to the digitization and automation of production processes, Industry 4.0 also has a significant impact on internal logistics, which is also gradually becoming a digital system thanks to the implementation and integration of technologies corresponding to Industry 4.0. Logistic support of production and modern distribution logistics have already been given a new name in the current debates about Industry 4.0, and there is talk of Logistics 4.0 or Smart Logistics [6]. The future challenges of internal logistics are primarily driven by growing individualization in all areas of the company and increasing customer requirements for logistics services. This creates the need for flexible and adaptable internal logistics systems as well a transparent and continuous exchange of information [7].

2. Literature review of the main topics for the methodology

The methodology proposed in this paper for the evaluation of internal logistics processes includes essential topics which are: internal logistics within the Industry 4.0 concept, logistics evaluation audit and readiness models for maturity assessment in Industry 4.0. Therefore, these key areas are subjected to literature research, and scientific methods such as analysis, synthesis, comparison and abstraction.

2.1. Internal Logistics and Logistics 4.0

The system of internal logistics is corporate logistics (supply logistics, production logistics and distribution logistics) [8]. Logistics can be divided according to the area of its application within the logistics chain, where internal logistics focuses on logistics processes within a company [9]. The primary operational functions of internal logistics are manipulation, warehousing, internal transport within a company (warehouse, production, etc.), picking and packaging [10]. The integration of new technologies into logistics promises to enable real-time monitoring of material flows, better storage and better handling of manipulation units [1]. Most authors describing logistics within the concept of Industry 4.0 identify and mention similar technologies and applications that are based on the principles of this concept and implemented in the field of internal logistics. The leading technologies of Logistics 4.0 are listed according to their sources in Table 1.

Table 1. Technologies of Logistics 4.0

Sources	Technologies of Logistics 4.0
Wang (2016) [11]	Big Data, Smart Sensors, RFID, IoT, IoS, Smart Robots, Autonomous AGV, GPS
Galindo (2016) [12]	RFID, CPS, IoT, IoS, Big Data, RTLS,
Schmidtke et al. (2018) [13]	Autonomous AGV, Mobile terminals, Cloud IT platforms, IoT, Data Mining
Wahrmann et al. (2019) [14]	Wearable Scanners (barcodes, RFID tags), Voice-Direct Headsets, Smart Glasses, Activity Trackers
Glistau and Machado (2018) [15]	RFID, WLAN, EDI, GPRS, Big Data, Data Mining, Smartphones, EDI, 5G network, Tablets, Cloud storage
Zou and Zhong (2018) [16]	Exoskeletons, Autonomous AGV, Drones, Collaborative Robots, RFID
Horenberg (2017) [6]	Autonomous AGV, Smart Robots, Blockchain Technology, CPS, Drones, Big Data

2.2. Logistics audit methodologies

This part describes the evaluation of logistics audit systems and methodologies, based on which the dimensions of internal logistics for the methodology are determined. Logistics audit is a standardized analytical and project process that focuses on the logistics functions of the corporate management system. The purpose of a logistics audit is to diagnose the state of the logistics system in the company and define a program to increase its productivity. The project nature of this program makes it possible to implement both immediate organizational interventions leading to rapid effects and fundamental conceptual changes in logistics processes [17]. The logistics audit aims to provide management with comparative material on the state of logistics in the company and to show the potential for changes in logistics activities in order to achieve a more efficient performance of the company [18]. Several methods of logistics audits were examined in detail, where the main parameters sought were the areas of internal logistics that the audit evaluates, as well as the potential for evaluating internal logistics corresponding to Industry 4.0. That is the integration of this concept into the logistics audit. Three critical methodologies and their main parameters were evaluated, which are shown in Table 2.

Table 2. Evaluation of logistic audits

Methodology by source	Evaluated areas in internal logistics	Integration of Industry 4.0	Field of application	Evaluation method
CHLA [19]	Material flow, Information system, Planning, Strategy	No	Universal	Point and subjective evaluation
VDA 6.3 [20]	Manipulation, Transport, Storage, Packaging	No	Automotive	Point evaluation
MMOG/LE [21]	Manipulation, Supply, Packaging, Material Identification	No	Automotive	Point evaluation

2.3. Readiness models for maturity assessment in Industry 4.0

Using readiness models, we can identify and evaluate the current state of the company in various areas within the concept of Industry 4.0. The individual models have standard foundations and very similar principles, which are based on CMMI models [22]. The analysed dimensions and levels are always defined. In some models, the dimensions are further extended to sub-dimensions for greater detail. In addition to dimensions, the method of evaluation is also essential. The similarity in the method of evaluation is given by individual degrees of maturity, readiness in given dimensions. There are usually four to six levels of readiness. The CMMI model has five levels from initial, controlled, defined, quantitatively controlled to optimizing [23]. Table 3 provides an overview of the detailed models analysed with the main parameters sought for use in the proposed methodology.

Table 3. Analysed models of readiness and their main parameters

Readiness model by source	Analysed dimensions/areas	Potential of the logistics	Method of assessing readiness	The purpose of the model
RM1 [24]	6 dimensions	Yes	6 degrees	Descriptive Comparative Prescriptive
RM2 [25]	Without dimensions	Yes	6 degrees	Comparative
RM3 [26]	5 dimensions	Yes	5 degrees	Descriptive
RM4 [27]	4 dimensions	Yes	5 degrees	Descriptive
RM5 [28]	7 dimensions	Yes	5 degrees	Descriptive
RM6 [29]	3 dimensions	Yes	5 degrees	Descriptive

Maturity models are arranged logically from the lowest level to the highest level. Degrees or levels have their names and their characteristics of requirements and properties for meeting a given degree within a given dimension. A total of 20 preparedness models were found, which evaluate companies within Industry 4.0 and their readiness for this concept. Within the methodology, we focus primarily on models that evaluate internal logistics. A selection was made in terms of quality and relevance, and models that contain the potential for evaluating internal logistics were further analysed.

2.4. Evaluation of literary research

The first part of the paper deals with the research of the main areas - internal logistics and Logistics 4.0, logistics evaluation audits and models of readiness for the concept of Industry 4.0. It is, therefore, possible to summarize the main findings:

- Based on the analysis and synthesis, we can state that the technical aspects of Logistics 4.0 in specific areas are based on technological assumptions and principles from the comprehensive general concept of Industry 4.0.
- The sources for Logistics 4.0 mention the principles and technologies that will be a necessary part of logistics processes in a company that will apply the attributes of Industry 4.0.
- After comparison and analysis, it was found that the logistics audit systems do not reflect the development within the concept of Industry 4.0 and do not evaluate the internal logistics processes with this regard.
- The activities of internal logistics, which evaluate some logistics audits, such as warehousing, material handling, packaging, supply, etc. are not evaluated in detail according to indicators.
- Based on a comparison of a company's readiness models for Industry 4.0, it is quite evident that internal logistics plays a minimal role here and the level of readiness is not assessed in detail
- According to the method of abstraction, it can be stated that for a smaller number of models directly conceived for logistics, the evaluation is performed in a general form instead for the whole area. Internal logistics occurs in the models as a separate evaluated area minimally

3. Methodology for evaluating the readiness of internal logistics processes

Based on the theoretical conclusions above and their related shortcomings, a new methodology is proposed to increase the readiness of internal logistics processes for Industry 4.0. It assesses all internal logistics processes and tries to structure them. Using the methodology, the processes are then evaluated, the current level is assessed, and modern elements are implemented in an effort to increase the level of the logistics processes. The methodology for evaluating the readiness of a company for Industry 4.0 is described in the analytical part, which has three main phases, which are also partial goals.

3.1. Definition and structure of the methodology in terms of company attributes

The methodology is designed with special consideration of the requirements of industrial enterprises. When dealing with production management systems, we must take into account the specifics of a company and its situation, as well as logistics systems. An adequately designed logistics system must make a positive contribution to the overall performance of a company and meet specific requirements in the industrial enterprise. Each company is also associated with certain features or characteristics that must be taken into account in any intervention in its internal logistics processes. These are, for example: the type of production and its repeatability, the size and type of the enterprise, the internal logistics system in the enterprise, logistics technology, data processing and information flow, and more. Table 4 summarizes the underlying assumptions and limitations of the proposed methodology.

Table 4. Summary of requirements and limitations of the methodology

Attribute	Assumption and limitations
Company size	Medium-sized companies
Branch of business	Manufacturing industrial enterprises

Production repeatability	It doesn't matter, it's not a limitation
Data availability	Available data from information systems, monitoring, control panels, etc.
Complexity of logistics processes	It doesn't matter, it's not a limitation
Primary focus	Functional and economical logistics management
Logistics technology	It doesn't matter, it's not a limitation

The analytical part of the proposed methodology is the subject of this practical study. Its phases are chosen according to a logical sequence. The aim of the analytical part is data collection, evaluation and identification of problem areas in the field of logistics. Indicators determining the effectiveness of the proposed methodology are also included. The analytical part itself takes place in three primary phases, and these are described below.

3.2. Analytical Phase 1 - Determination of internal logistics dimensions, levels and indicators for evaluation

Analytical phase 1 includes the main steps, such as defining the critical dimensions of the internal logistics system, setting the levels for evaluation and setting the evaluation indicators of the individual dimensions of internal logistics.

3.2.1. Step 1 - Determination of dimensions The dimensions of the internal logistics are determined using the methods of logistics audits according to MMOG/LE [21], VDA 6.3 [20] and the Chamber of Logistics Audits (CHLA) [19]. The specified dimensions and sub-dimensions cover all the internal logistics processes. The five main dimensions of the methodology are: *Manipulation Storage Supply Packaging Material identification*

These 5 main dimensions, which were selected for the methodology as the main activities of internal logistics, must be further broken down into sub-dimensions, because the dimensions are very general and evaluating it as a whole area is inefficient and meaningless.

Manipulation dimensions – The material is moved from one place to another, and an information change of location is performed. This is the manipulation between the receipt and storage of material, the production sector and the dispatch of finished products. Specified sub-dimensions:

- Manipulation technology
- Information security during manipulation
- Manipulation unit

Storage dimensions – Loading the material and storing it, entering it in the information system where the material is stored, the method of requesting the material from the warehouse, physical removal, preparation for removal and again the information entry into the system. Specified sub-dimensions:

- Storage technology
- Information security during storage
- Material receipt and shipping

Supply dimension – What is important is the technique used for loading the material (randomly, intervals, Milk Run), how the material is required (paper order, kanban, EDI, electronic system, a system of empty packages and free positions, etc.). Specified subdimensions:

- Workplace supply technology
- Method of material delivery
- Information about material recalls

Packaging dimensions – The packaging is made of different materials, and it has different functions (protection, breathability, etc.). Packing can be done manually in a carton with insulating tape, or using various tools and technologies, or the process is fully automated and robotic. Specified sub-dimensions*:*

- Packaging technology, method of packaging

- Types of packaging and packaging material
- Environment and packaging management

Material identification dimensions – It is a process of placing an identifier and scanning, and loading into the system. Interconnected production and storage systems must know precisely where, when and in what quantity a particular component is to enable this modern approach to production. In internal logistics, therefore, the most important thing for traceability is a well-set system of material marking and information acquisition within the entire internal logistics flow. Specified sub-dimensions:

- Method of identification
- Information security and identification intensity

3.2.2. Step 2 - Levels of evaluation of logistics processes The level of readiness of internal logistics processes from Industry 4.0 is divided into six levels based on analysis and synthesis. Therefore, six levels of preparedness, including the zero level, are set for the methodology of evaluating all the sub-dimensions as follows:

- Level 0 - Processes are not explicitly defined
- Level 1 - Certified process management
- Level 2 - Digitized data collection (from processes), implementation of automation
- Level 3 - Part of the processes is automated, connected to an external data source
- Level 4 - Processes are automated, with limited human intervention
- Level 5 - Processes are automated, only human-controlled

The levels are evolving, and the brief definition only gives a general idea about what each level means. However, for the evaluation within the methodology, the levels are characterized in even more detail.

3.2.3. Step 3 - Intersection of dimension and level parameters and links between levels In this step, links between the individual levels are first created, and they are used to create the intersection of the parameters of the dimension (sub-dimension) and the level of readiness.

The subject of the third step is to determine:

- Links between individual levels, a transition to a higher level
- Define and describe the intersection of these two parameters, a brief description
- Assign for a given subdimension and level, in addition to the characteristic, also the logistic elements used in the given level

An example of the characteristics of the fourth and fifth levels, including the assigned logistics elements, is shown in Table 5. These levels correspond to Industry 4.0. This is the dimension of manipulation and its sub-dimension of manipulation technology. However, this characteristic is performed for all levels, including elements.

Table 5. Demonstration and characteristics of the subdimension of manipulation technology

Level	Characteristics of the subdimension	Logistics elements
4	Manipulation technology is automated, controlled and guided by IS. A person loads and unloads material. There is no autonomous solution, and there is a gradual robotization of workplaces. Automated manipulation technology is only used for material transport.	Automated manipulation equipment - forklifts, trains, trucks guided by sensors built into the ground. Collaborative robots.

5	Manipulation technology is fully automated, controlled and guided by IS. The human role is only as overseer. Artificial intelligence is used - a robot loads and unloads material. Manipulation technology connects the line and the warehouse autonomously, there is a link to robotic workplaces.	Automated manipulation equipment - unmanned forklift, autonomous forklift, trains, drones. The robot loads and unloads material. Autonomous AGV.

3.2.4. Step 4 - Determination of indicators Indicators are used to evaluate the sub-dimensions. These are of a general type, which is of a qualitative nature, and their task is to find out in more detail the type and condition of the company and the initial information. Another type are evaluation indicators, which are quantifiable in nature and are at the core of determining the level of internal logistics in a company. Each of the defined sub-dimensions contains 2-3 quantifiable indicators, which by their nature, provide a complete picture of the area. For example, the indicators in Table 6 are evaluated for the package dimension or its sub-dimension.

Table 6. Indicators for packaging sub-dimensions

Packaging	**Indicators**		
Packaging technology	Degree of automation	Technical equipment	Customizable product size
Types of packaging	Packaging identification	Packaging standardization	-
Environment and packaging management	Applicability (repeatability)	Attitudes towards packaging management	Carbon footprint

An important note for understanding the principle is that these indicators will then be used in a structured questionnaire and, based on the answers and evaluation, the company will be classified at a certain level according to the indicators (0 - 5). The individual levels will be divided as a percentage for each indicator.

3.3. Analytical Phase 2 - Evaluation tool and structured interview

Phase 2 of the analytical part of the methodology is focused on conducting a structured interview with employees of the company. In phase 2, the data created in phase 1 of the same part of the methodology are needed. This is an essential part of the proposed methodology, as the results of the evaluation largely depend on the quality of the structured interview. The questionnaire is designed to match the levels and indicators. Thus, a total of 38 questions are created, and the answers are in the form of brief sentences, a percentage, or a multiple-choice style question. The questions are designed so that not only one evaluation body answers them.

3.4. Analytical Phase 3 - Final evaluation of the level of internal logistics

The last phase of the analytical part aims to determine the final evaluation of each sub-dimension, the evaluation of each parent dimension and the overall level of internal logistics in the company. In this phase, the answers are evaluated using the scales for each indicator. The essence of determining the level of internal logistics is multi-criteria evaluation, i.e. evaluation using criterion functions. A responsible employee at the company is asked questions related to a specific indicator in the given area. Each indicator is always evaluated by levels from level 0 to level 5. The employee then chooses from the above options and does not know how many points they will receive for a specific answer. This is carried out in the same way for all 38 indicators. The output of the analytical phase is an evaluated, structured interview with a clear score in each indicator. The individual dimensions of logistics are interlinked, and the output is an assessment of the current level of internal logistics, and the company will discover the areas where its level can be improved and increased.

3.5. The current state of methodology and its future extension - implementation phase

Currently, the newly proposed methodology is being verified in companies. The output is an assessment of the current level of internal logistics. The methodology is therefore descriptive, and the company will find out the areas where there is the potential to improve and increase its level. This means that the company knows where to improve, but does not know the implementation steps to achieve it. Therefore, the methodology can be extended by an implementation part in the future. The result of phase 3 of the analytical part is also the input to the implementation part. At the very beginning of the implementation phase, these outputs are further processed into the values of the levels of individual sub-dimensions or indicators. The implementation phase deals with the design of corrective measures aimed at improving the situation in the identified areas. The implementation phase can be divided into three key steps. The first steps include processing the outputs from the analytical part and a detailed overview of individual areas. The next step is the definition and scope of measures to achieve a higher level in the area, including the necessary methods and tools. The third step is the implementation of the measures.

4. Conclusions

Internal logistics is an essential part of factory production processes and an integral part of the supply chain. Internal logistics is a suitable area for applying the principles of Industry 4.0. The vision of smart logistics is to connect warehouse systems, internal logistics, transportation technology and planning software into one interconnected network that can automatically create and modify logistics processes related to other links in the production chain. Digitized internal logistics ensures the smooth automated transfer of materials, goods and machines within a company - between warehouses and the production area. The field of internal logistics has many possibilities and great potential within the concept of Industry 4.0. We decided to exploit this potential by creating a missing methodology for the evaluation of internal logistics processes in companies. To be able to implement new technologies and raise and improve a company's level, it is necessary to know its current level and the logistics processes themselves. The methodology, therefore, contains an analytical part divided into the main steps for determining the current level of internal logistics in the company in individual dimensions. Using the criterion function it is possible to link these dimensions and determine the overall level of internal logistics of the company. This methodology can, therefore, be used as a diagnostic tool. The methodology also has the potential for expansion, while an implementation section can be added to the analytical part and the methodology will become more comprehensive. After evaluating its current level, a company implements measures to increase the level of any of its inadequate dimensions. Verification of the methodology and results from companies will be the subject of another article, as well as a possible extension of the methodology with an implementation section.

Acknowledgments

This research was supported by the project SGS-2018-031 Optimizing the parameters of a sustainable production system.

References

[1] Hofmann E and Rüsch M (2017). Industry 4.0 and the current status as well as future prospects on logistics. *Computers in industry*, **89**, 23-34.

[2] Mařík V (2016). *Průmysl 4.0: výzva pro Českou republiku*. Management Press.

[3] Kagermann H Helbig J Hellinger A and Wahlster W (2013). Recommendations for implementing the strategic initiative INDUSTRIE 4.0: Securing the future of German manufacturing industry; *final report of the Industrie 4.0 Working Group*. Forschungsunion.

[4] Schmidt R Möhring M Härting R C Reichstein C Neumaier P and Jozinović P (2015, June). Industry 4.0-potentials for creating smart products: empirical research results. In *International Conference on Business Information Systems* (pp. 16-27). Springer, Cham.

[5] Rüßmann M Lorenz M Gerbert P Waldner M Justus J Engel P and Harnisch M (2015). Industry 4.0: The future of productivity and growth in manufacturing industries. *Boston Consulting Group*, **9(1)**, 54-89.

[6] Horenberg D (2017). Applications within Logistics 4.0: A research conducted on the visions of 3PL service providers (Bachelor's thesis, University of Twente).
[7] Günthner W and Ten Hompel M (2010). *Internet der Dinge in der Intralogistik* (p. 360). Heidelberg: Springer.
[8] Sixta J and Mačát V (2005). *Logistika: teorie a praxe.* CP Books, as.
[9] Stehlík A *Logistika – strategický factor manažerského úspěchu.* 1. vyd. Brno : Studio Contrast, 2002. 231 s. ISBN-80-238-8332-1
[10] Martin H (2012). *Transport-und Lagerlogistik: Planung, Struktur, Steuerung und Kosten von Systemen der Intralogistik.* Springer-Verlag.
[11] Wang K (2016, November). Logistics 4.0 Solution-New Challenges and Opportunities. In *6th International Workshop of Advanced Manufacturing and Automation.* Atlantis Press.
[12] Domingo Galindo L (2016). *The challenges of logistics 4.0 for the supply chain management and the information technology* (Master's thesis, NTNU)
[13] Schmidtke N Behrendt F Thater L and Meixner S (2018). Technical potentials and challenges within internal logistics 4.0. In *2018 4th International Conference on Logistics Operations Management (GOL)* (pp. 1-10). IEEE.
[14] Wahrmann D Hildebrandt A C Schuetz C Wittmann R and Rixen D (2019). An autonomous and flexible robotic framework for logistics applications. *Journal of Intelligent & Robotic Systems,* **93(3-4)**, 419-431.
[15] Glistau E and Coello Machado N I (2018). Industry 4.0, logistics 4.0 and materials-Chances and solutions. In *Materials Science Forum* (Vol. **919**, pp. 307-314). Trans Tech Publications.
[16] Zou O and Zhong R Y (2018, July). Automatic Logistics in a Smart Factory using RFID-enabled AGVs. In *2018 IEEE/ASME International Conference on Advanced Intelligent Mechatronics (AIM)* (pp. 822-826). IEEE.
[17] Komora Logistických Auditorů. Nová metodika logistického auditu. *Ostrava: KLA*
[18] Malindžák D and Marková Z (2010) Systém komplexného logistického auditu. **15** (2010), mimoriadne číslo 1. *Acta Montanistica Slovaca*, s. 82-89.
[19] Tvrdoň L Bazala J and David R (2012). A new approach of the logistics audit. In CLC 2012: (pp. 66-71).
[20] der Automobilindustrie, V. V. (2009). *Qualitätsmanagement in der Automobilindustrie Bd, 4: Sicherung der Qualität in der Prozesslandschaft.* VDA.
[21] Peinado J and Graeml A R (2008). Mensuração do desempenho logístico: a percepção da eficácia da ferramenta MMOG-LE por fornecedores da indústria automobilística. *Simpósio de Administração da Produção, Logística e Operações Industriais (Simpoi),* **11**.
[22] CMMI, P. T. (2010). CMMI for Development, Version 1.3. *Software Engineering Institute.*
[23] Paulk M C Curtis B Chrissis M B and Weber CH V. Capability maturity model, version 1.1. *IEEE software 10*, no. **4** (1993): 18-27.
[24] Lichtblau K Stich V Bertenrath R Blum M Bleider M Millack A and Schröter M (2015). IMPULS-industrie 4.0-readiness. *Impuls-Stiftung des VDMA, Aachen-Köln.*
[25] Werner-Lewandowska K and Kosacka-Olejnik M (2019). Logistics 4.0 Maturity in Service Industry: Empirical Research Results. *Procedia Manufacturing*, **38**, 1058-1065.
[26] De Carolis A Macchi M Kulvatunyou B Brundage M P and Terzi S (2017, July). Maturity models and tools for enabling smart manufacturing systems: comparison and reflections for future developments. In *Ifip international conference on product lifecycle management* (pp. 23-35). Springer, Cham.
[27] Sternad M Lerher T and Gajšek B (2018). Maturity levels for logistics 4.0 based on NRW's Industry 4.0 maturity model. *Business Logistics in Modern Management.*
[28] Krajcovic M Grznar P Fusko M and Skokan R (2018). Intelligent Logistics for Intelligent Production Systems. *Communications-Scientific letters of the University of Zilina,* **20(4)**, 16-23.
[29] Oleśków-Szłapka, J and Stachowiak A (2018, September). The framework of logistics 4.0 maturity model. In *International Conference on Intelligent Systems in Production Engineering and Maintenance* (pp. 771-781). Springer, Cham.

IOP Conf. Series: Materials Science and Engineering **947** (2020) 012006 doi:10.1088/1757-899X/947/1/012006

Tracking Primary Dendritic Arm Evolution through Single Crystal CMSX-10® Ni-Base Superalloy

J Strickland[1]*, B Nenchev[1], S Perry[1], K Tassenberg[1], HB Dong[1]

[1]School of Engineering, University of Leicester, Leicester, LE1 7RH, UK

jcjs2@leicester.ac.uk

Abstract. A novel packing pattern and local primary dendrite arm spacing (PDAS) algorithm, Shape-Limited Primary Spacing (SLPS), is applied to single crystal CMSX-10® Ni-base superalloy. The characterisation technique is rapid, repeatable and comparable, enabling the standardisation of single crystal microstructure analysis. It is deduced, that the increase in local PDAS range observed within the last to solidify sections of the CMSX-10® sample used this study, is due to lateral macrosegregation. SLPS enables quantification of the influence of the processing parameters on resultant packing patterns and local PDAS distribution. Through refinement of local PDAS, deleterious non-hexagonal pattern formation can be reduced. SLPS permits the optimisation of single crystal microstructure and the development of superior quality components.

1. Introduction

Single crystal alloys find a wide range of applications, from semi-conductors, optoelectronics to applications in aerospace engines. Single crystals components used in aerospace are manufactured via the Bridgman directional solidification process, with a seed or grain selector at the bottom the casting [1]. The specific casting conditions lead to the formation of organised columnar arrays of dendrites whose [001] crystal direction is well-aligned with the withdrawal direction. On a 2D etched sample section perpendicular to the growth direction, dendrites appear as bright cross-like features surrounded by dark interdendritic material.

The formation of the dendrites within the array is a result of thermo-physical processes that occur in the mushy zone during solidification [2]. It has been suggested that single crystal mosaicity is caused by plastic deformation in the mush which results in crystal lattice rotations and dendrite misorientation [3-5]. The resultant, low-angle grain boundaries (LABs) are detrimental to the creep rupture life, especially when the local misorientation angle is larger than 9° [6].

The primary dendrite arm spacing (PDAS) characterises the distance between neighbouring dendrites. PDAS is intrinsically linked to the solidification environment, local misorientation, and indicates the maximum length scale for microsegregation [7]. Recently a new single crystal characterisation method, Shape-Limited Primary Spacing (SLPS) [8], has been developed which can automatically, accurately, and rapidly determine the local PDAS distribution within a single crystal bulk array. SLPS can accurately determine the relationship between dendritic packing arrangements, local PDAS, and tip growth kinetics. Consequently, it is now possible to compare multiple single crystal cross-sections and relate variations in processing conditions to changes in the microstructure.

The aim of this work is to apply the SLPS methodology to a multiple sections of a single crystal bulk cast and demonstrate its applicability for process and alloy optimisation. The end goal of this new

IOP Conf. Series: Materials Science and Engineering **947** (2020) 012006 doi:10.1088/1757-899X/947/1/012006

approach is to improve understanding of the influence of the solidification conditions on single crystal growth and create a standardised methodology for characterisation.

2. Method

A single crystal CMSX-10® bar of 9.4 mm diameter and 64 mm in height (figure 1(b)) was provided by Rolls-Royce Plc. The bar was sectioned into seven segments perpendicular to the long axes of the sample using a metallographic precision cutter and mounted in Conducto-Mount. Section 1 corresponds to the part of the sample that solidified nearest to the chill plate of the Bridgman furnace. The bar was formed in a cluster mould arrangement and withdrawn from the Bridgman furnace at a constant withdrawal velocity, V. The sectioned samples were prepared for Scanning Electron Microscopy (SEM) using a standard grinding and polishing metallographic procedure and cleaned with CH_3OH between each polishing step to a 1 μm finish. At the end, 10 minute ultra-sonication removed excess diamond particles which agglomerate in regions of porosity. Figure 2(a) was acquired with an FEI 650 Quanta SEM at the University of Leicester, with a 20kV beam, WD = 10.1 mm, HF = 2.32 mm. The samples were imaged in back scattered electron (BSE) mode to visualise the microsegregation between the dendritic core and interdendritic region. MAPS® software (ThermoFisher Scientific, Waltham, MA, USA) provided large, high resolution, stitched images of the dendritic arrays automatically, with only minor adjustments required where necessary. The final images were comprised of a 24 × 24 grid of individual SEM images taken at 300 × magnification with 20 μs scan speed. The BSE image are loaded into the DenMap software [9] and the exact dendritic core pixel positions were located. These coordinates were input into the SLPS algorithm [8]. The dendritic packing and local PDAS distributions were calculated by analysing each individual dendrite in the array and determining true nearest neighbour (TNN) dendrites. TNN are defined as dendrites whose diffusion fields interact locally with a central dendrite; determined by the SLPS algorithm. This method of TNN determination is applied to each dendrite within the single crystal array. The total time for analysis for each sample section was 1 min and 20 seconds. Only the section 1 image (figure 1(b)) is included in this work to demonstrate the resultant dendritic mapping (figure 2(a)) and packing pattern determination (figure 2(b)).

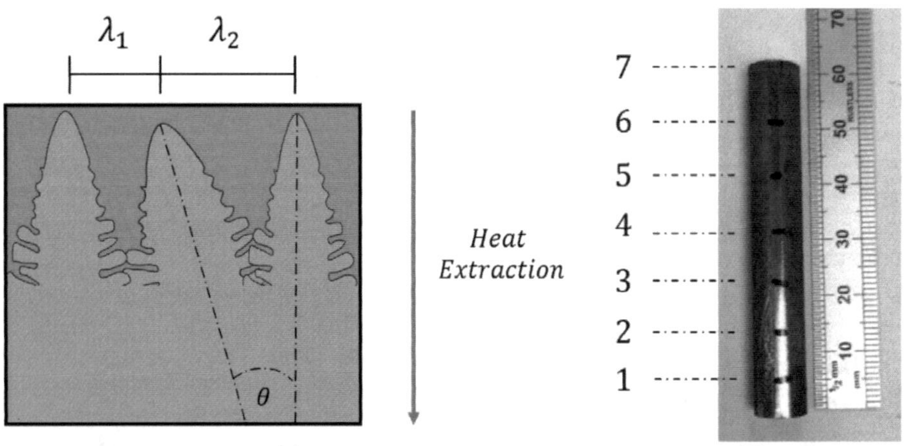

(a) (b)

Figure 1. (a) Local misorientation, θ, characterised by variation in local PDAS. The fastest growth direction is always the [001] direction aligned with the dendrite stem. A misorientated dendrite is outcompeted by its neighbours who are better orientated with the heat flow direction. (b) 64 mm CMSX-10® single crystal bar sample, $\emptyset = 9.4\ mm$. Sample section 1 is closest to the chill plate.

3. Results

The variation in determined packing patterns for sample section 1 are illustrated in figure 2(b). The centre of the sample demonstrates clear hexagonal packing (N_6). The outside edge of the sample a combination of triangular (N_3), square (N_4), pentagonal (N_5), heptagonal (N_7), octagonal (N_8) and nonagonal (N_9) packing. The total normalised packing obtained within each sample section is illustrated in Figure 2(c)). The non-dimensionalised histogram enables the comparison of all sample sections, under any change in solidification condition, regardless of geometry. The N_6 packing is the most frequently observed packing pattern for all sample sections (figure 2(c)). Between 1 and 7 there is a 9% combined increase in N_7 and N_8, a 5% combined decrease in N_3, N_4 and N_5 and a 4% increase in N_6 packing patterns. The distribution of local PDAS for all sample sections from figure 2(b) is shown in figure 2(d). The smallest PDAS for each sample section is determined as roughly 150 μm. The mean PDAS, $\bar{\lambda}$, and standard deviation, λ_σ, are the smallest for section 1. Over the first 50 mm of the sample there is very little change in the $\bar{\lambda}$ and λ_σ, the increase is almost linear with height from the bottom of the sample. However, samples 6 – 7 show large increases in the $\bar{\lambda}$ and λ_σ no longer follow the same relationship. The largest PDAS are observed within sample section 7.

The averaged lower and upper PDAS bounds for each packing pattern for sample section 1 and 7 are shown in figure 2(e). The results demonstrate that the average PDAS increases with coordination number, N. Within the single crystal microstructure N_7 and N_8 patterns are the largest observed arrangements and N_3, N_4 and N_5 the smallest. The lower PDAS bound for all sample sections is similar (figure 2(d)). In addition, the increase in PDAS per N for the lower bound demonstrates negligible change throughout the cast height (figure 2(e)). Interestingly, the upper PDAS bound for section 7 is significantly larger than that for section 1 (figure 2(e)). The range in PDAS for the N_3 patterns is only slightly greater for sample section 7. However, there is a significant increase in PDAS range for the N_8 patterns in sample section 7.

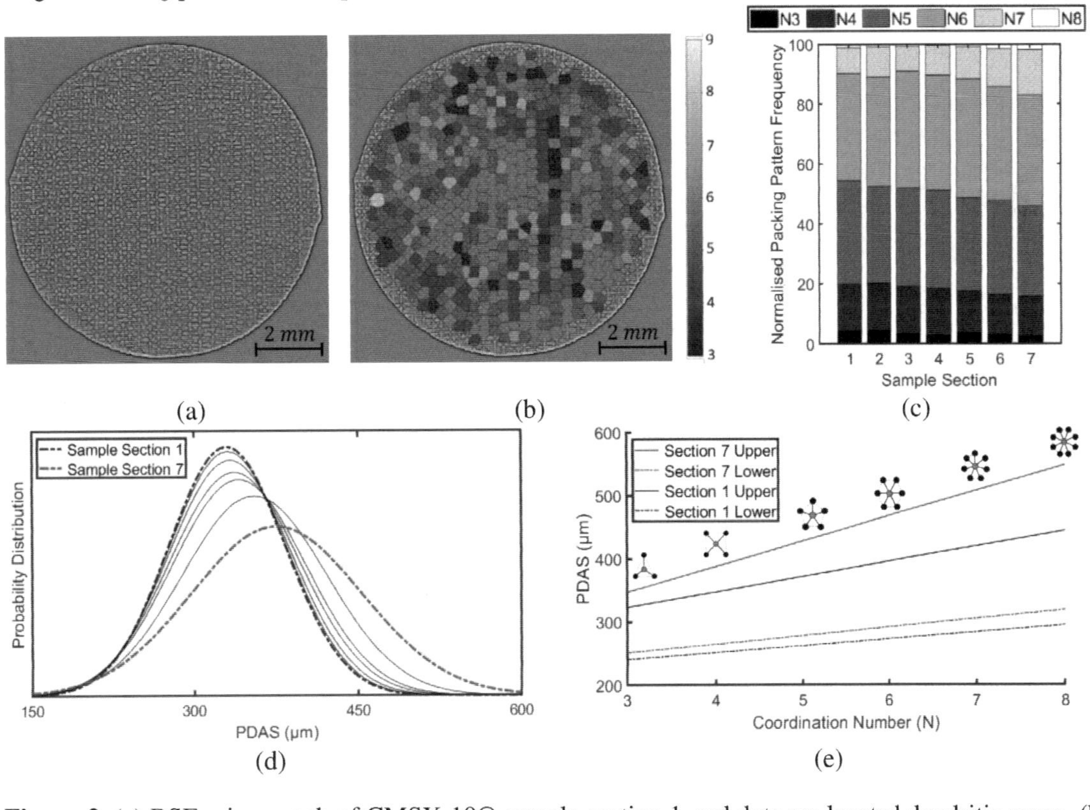

Figure 2. (a) BSE micrograph of CMSX-10® sample section 1, red dots are located dendritic cores. (b) Shape-Limited Primary Spacing (SLPS) dendritic packing analysis applied to the microstructure to determine local PDAS variation and packing pattern quantity. (c) Normalised stacked bar plot of all

determined packing patterns within each sample section from figure 1(b). (d) PDAS probability distribution for all 7 CMSX-10® sample sections, local PDAS variation is determined using Shape-Limited Primary Spacing (SLPS). (e) The variation in the mean lower and upper PDAS bounds for each coordination number for sample section 1 and 7.

4. Discussion

The PDAS is known to follow the steady-state non-linear relationship $\lambda = AC_0^{0.25}V^{-0.25}G^{-0.5}$ [10-16], where, G is thermal gradient, V is withdrawal velocity, C_0 is initial composition and A is a materials constant. Therefore, any variation in C_0 or solidification rate, $G \times V$, will result in a change in the PDAS. Consequently, the increase in $\bar{\lambda}$ and λ_σ observed between sample section 1 and 7 (figure 2(d)) is attributed to these two variables. In Bridgman casting, the axial G reduces with height from the chill plate as the fraction of solid superalloy increases [17]. The lower part of a mould is mainly cooled by intensive interaction with the chill plate, resulting in a solid/liquid isotherm perpendicular to the withdrawal direction. In a cluster mould arrangement, the temperature profile across the sample is uneven [18] and the isotherm typically attains a concave shape. Any isotherm curvature causes a variation in the macroscopic solid/liquid interface shape across the sample cross-section. As a result, radial convection is induced by temperature and thus density differences within the melt. Any spatial density difference causes movement of solute across the sample surface, directly influencing the liquid compositional gradient within the over lying melt parallel to the growing solid/liquid interface. According the steady state PDAS equation, solute enriched regions will experience an increase in PDAS (for solute partitioning to the liquid, $k < 1$), and solute depleted regions (where, $k < 1$) the PDAS decreases. The lower PDAS bound in figure 2(e) increases only slightly between section 1 and 7, therefore, the solidification environment ($G \times V$ and C) within a part of each bulk sample section, remained roughly constant throughout the cast height. Interestingly, the upper PDAS bound increases monotonically with sample section height (figure 2(e)).

The areas with largest PDAS are the N_7 and N_8 packing patterns (figure 2(c)), therefore, the increase in local PDAS range within the end sections of the bar (figure 2(e)) is intrinsically associated to these patterns. In CMSX-10®, N_7 and N_8 arrangements indicate areas rich in interdendritic partitioning elements such as Ni, Ti, Ta, and Al [19]. As shown in figure 2(b), N_7 and N_8 dendrites are almost always surrounded by smaller N_3, N_4 and N_5 arrangements. The N_3, N_4 and N_5 patterns indicate regions of dense packing; thus, areas rich in dendritic core elements [20]. The N_6 packing arrangement is the most frequently observed within all sample cross-sections (figure 2(c)). It is located between the N_3 and N_8 PDAS extremes (figure 2(d)) and achieves a $\bar{\lambda}_{N_6}$ (figure 2(e)) almost equal to the $\bar{\lambda}_{Array}$ (figure 2(d)). Regular N_6 hexagonal packing is the only arrangement that achieves a uniform PDAS between all TNNs. Therefore, if N_3, N_4 and N_5 are core rich and N_7 and N_8 interdendritic rich, N_6 indicates regions growing with equal microsegregation between neighbours, and with no local misorientation in respect to the heat flow direction. Consequently, arrangements of non-hexagonal packing indicate the location of competitive growth, where core rich N_3, N_4 and N_5 dendrites are outcompeting slower growth rate N_7 and N_8 neighbours (figure 1(a)).

As solidification height from the chill plate increases, dendrites eject interdendritic material into the liquid. For solutally unstable alloys such as CMSX-10®, solute enrichment occurs in the melt due to the slow diffusion speed in the solid superalloy. Between sample section 1 and 7 this enrichment corresponds to a 9% increase in solute enriched N_7 and N_8 dendrites and a 5% decrease in solute depleted N_3, N_4 and N_5 (figure 2(c)). If the solute concentration parallel to the growing interface is uniform, then according to the steady state PDAS equation, a change in bulk G, V, or C only influences the $\bar{\lambda}_{Array}$. Changing the bulk solidification conditions plays no role in determining the local PDAS range. Therefore, as the $G \times V$ across a bulk sample perpendicular to the withdrawal direction is constant (assuming negligible change in the isotherm shape), then the increase in local PDAS range in the end sections of the CMSX-10® cast (figure 2(d)) must be a result of increasing lateral solute enrichment (for $k < 1$).

Using the SLPS algorithm, it is now possible to understand the relationship between the bulk solidification environment, packing patterns, local PDAS, and tip growth kinetics. By tailoring the processing parameters and/or alloy composition the bulk local PDAS range can be refined. Through application of SLPS the bulk hexagonality of a single crystal array can be increased, thus, decreasing the formation of deleterious non-hexagonal packing patterns. Consequently, the SLPS methodology permits the development of superior quality single crystal components.

5. Conclusions

A novel packing pattern and local primary dendrite arm spacing (PDAS) algorithm, Shape-Limited Primary Spacing (SLPS), is applied to study a CMSX-10® single crystal Ni-base superalloy. It is deduced, that the broadening in local PDAS range observed within the last to solidify sections of a single crystal cast is a result of lateral macrosegregation. The new packing pattern characterisation method is rapid, repeatable, and accurate. SLPS enables quantification of the influence of the processing parameters on resultant packing patterns and local PDAS distribution within an array. Through application of SLPS and the tailoring of the processing parameters and/or alloy composition, the PDAS can be refined, reducing the severity non-hexagonal packing formation; thus, improving the quality of single crystal components.

Acknowledgements

We gratefully acknowledge the financial support from the Centre for Doctoral Training in Innovative Metal Processing (IMPaCT) funded by the UK Engineering and Physical Sciences Research Council (EPRSC) and Rolls-Royce Plc for providing financial support and supplying the single crystal CMSX-10® sample.

6. References

[1] Ma D 2018 *Front. Mech. Eng.* **13** 3-16
[2] Kurz W, Fisher DJ, Trivedi R 2019 *Int. Mat. Rev.* **64** 311-54
[3] D'Souza N, Newell M, Devendra K, Jennings P, Ardakani M, Shollock B 2005 *Mat. Sci. Eng. A* **413** 567-70
[4] Newell M, Devendra K, Jennings P, D'Souza N 2005 *Mat. Sci. Eng. A* **412** 307-15
[5] Newell M, D'Souza N, Green N 2009 *Int. J. Cast Metal. Res.* **22** 66-9
[6] Siredey N, Boufoussi MB, Denis S, Lacaze J 1993 *J. Cryst. Growth* **130** 132-46
[7] Hunt J, Lu S-Z 1996 *Metall. Mat. Trans. A* **27** 611-23
[8] Strickland J, Nenchev B, Dong HB, 2020 *MCWASP* **861** 012057
[9] Nenchev B , Strickland J, Tassenberg K , Perry S, Gill S, Dong HB 2020 *J. Imaging* **6** 19
[10] Kurz W, Fisher D 1981 *Acta Metall* **29** 11-20
[11] Trivedi R 1980 *J. Cryst. Growth* **49** 219-32
[12] Han SH, Trivedi R 1994 *Acta Metall. Mater* **42** 25-41
[13] Gandin CA, Eshelman M, Trivedi R 1996 *Metall. Mater. Trans. A* **27** 2727-39
[14] Ma D, Sahm PR 1998 *Metall. Mater. Trans. A* **29** 1113-9
[15] Bouchard D, Kirkaldy JS 1997 *Metall. Mater. Trans.s B* **28** 651-63
[16] Trivedi R 1984 *Metall. Mater. Trans. A* **15** 977-82
[17] Szeliga D, Kubiak K, Motyka M, Sieniawski J 2016 *Vacuum* **131** 327-42
[18] Lian Y, Li D, Zhang K 2015 *J. Cryst. Growth* **426** 186-97
[19] Bogdanowicz W, Albrecht R, Sieniawski J, Kubiak K 2014 *J.l Cryst. Growth* **401** 418-22
[20] Caldwell E, Fela F, Fuchs G 2004 *JOM* **56** 44-8

IOP Conf. Series: Materials Science and Engineering **947** (2020) 012007 doi:10.1088/1757-899X/947/1/012007

Comprehensive Quantitative Characterisation of Single Crystal Alloys

B Nenchev[1*], J Strickland[1], S Perry[1], K Tassenberg[1], S Gill[1]

[1]School of Engineering, University of Leicester, Leicester, LE1 7RH, UK

bn55@leicester.ac.uk

Abstract. Classically, the primary dendrite arm spacing (PDAS) is used as a gauge to determine the quality of a single-crystal alloy. However, this length scale varies substantially over a cast due to variation in the thermo-solutal conditions. In this work, several dimensionless parameters are derived that characterise this variation based purely on geometrical and natural pattern formation. They are determined by applying a novel shape limited spacing algorithm (SLPS). SLPS accurately evaluates the relationship between local PDAS and packing pattern formations, offering a new method of single crystal analysis. The SLPS algorithm is a part of DenMap, a dendrite core feature extraction tool designed to significantly speed up single-crystal characterisation for process optimisation and standardisation for academia and industry.

1. Introduction

Single-crystal turbine blades are the most important part of a gas turbine engine. They are the limiting factor in determining the turbine entry temperature (TET) which subsequently controls the operating efficiency of the engine. They are manufactured by a complex and expensive unidirectional solidification process, which aligns the [001] crystallographic plane with the direction of heat flow. In directionally solidified materials, for a given composition, the growth rate and temperature gradient determine the morphology of the microstructure and the level of microsegregation [1, 2]. Large primary dendrite arm spacings (PDAS) indicate increased levels of microsegregation, which negatively impacts the high temperature performance of the single crystal material. To reduce microsegregation severity, solution heat treatment is applied to homogenise the microstructure; the time and temperature of which is estimated on the maximum PDAS observed. Therefore, improving the predictive capability of material processing–structure–property relationship models relies upon developing an enhanced understanding of single crystal growth formation and accurate quantitative stereological characterisation of the as-cast microstructure.

Intensive manual microstructural analysis is performed in industry and research worldwide to carry out single crystal alloy characterisation and furnace calibrations. A typical data set can contain 10 or more serial slices with at least 300 dendrite cores per slice, all of which require manual identification [3]. This overwhelming amount of data, and thus time, causes a bottleneck at the image processing, core mapping and characterisation step. To understand the role of PDAS on single crystal performance, design future alloys and improve processing routes, a confluence of automated and standardised characterisation techniques are required. Currently, no universal method for single crystal characterisation exists, especially one that incorporates comparison between different single crystal alloys formed under a multitude of conditions.

IOP Conf. Series: Materials Science and Engineering **947** (2020) 012007 doi:10.1088/1757-899X/947/1/012007

In this work, a method for big data analysis and characterisation standardisation is proposed with the goal of developing a universal database for all single crystal alloys. The eventual aim is to link processing conditions to properties in a way that enables development of machine learning algorithms and computational models to optimise single crystal manufacture and the design of new alloys.

2. Automated characterisation

For automisation of single crystal characterisation, the DenMap software tool was developed. The complete DenMap software package is shown in figure 1. It uses an automated feature extraction algorithm designed for rapid, accurate and repeatable dendrite core detection [4] . The algorithm implements a Normalised Cross-Correlation (NCC) method which performs a comparison between a template image and an image of interest. For NCC to be effective, the dendritic cores must be clearly visible and possess good definition. This makes the DenMap algorithm ideal for SEM and optical micrographs as there is a clear contrast variation between the dendritic and interdendritic regions due to the preferential microsegregation of lighter and heavier elements, as shown in figure 2(a). A 600 dendrite micrograph can be automatically processed in less than 90 seconds with a dendrite core detection accuracy of 98%, figure 2(b). The automatic process includes mount and noise removal, image filtering via a customised Fast Fourier Band Pass filter (FFT Bandpass), template generation, thresholding and core mapping. Once the core positions are obtained, a subsequent stereological characterisation step is performed.

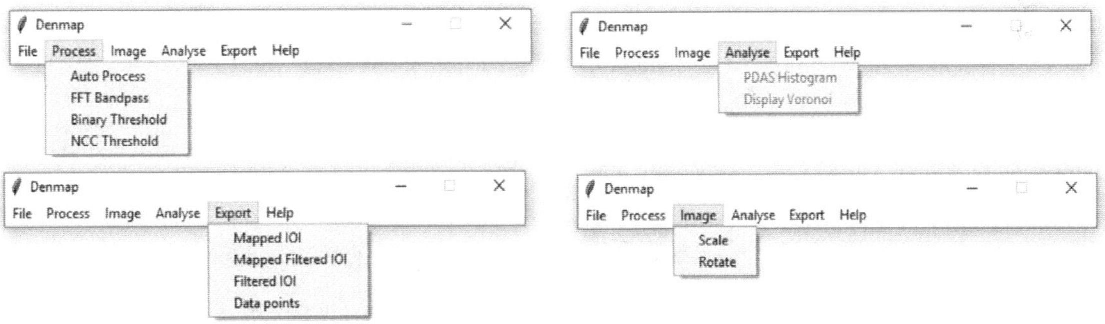

Figure 1. The DenMap software package toolbar, showing the available options for images analysis, extraction and statistical data acquisition.

Characterisation is achieved by a fully automatic and accurate shape limited spacing algorithm (SLPS) [5]. The software performs a Voronoi analysis to determine the maximum possible dendritic nearest neighbours. Then a systematic and consistent packing algorithm, utilising a local polygon shape factor (K_{SLPS}), is applied to determine the true nearest neighbours (TNN) i.e. determine only those dendrites that are affecting each other. For dendritic and cellular alloys, the algorithm successfully lowers the upper PDAS bound and returns a perfect Gaussian distribution of local PDAS. This accurate and repeatable single crystal characterisation has never been achieved before [6-8]. The dimensionless polygon shape factor, K_{SLPS}, is derived from the square counting method [9] as follows:

$$K_{SLPS} = \sqrt{\frac{N}{A_{TNN}}} \quad (1)$$

where A_{TNN} is the area bounded by the TNN, $\bar{\lambda}$ is the local PDAS and N is the number of TNN. An illustrative example of the DenMap algorithm is shown in figure 2, where a CMSX-4® micrograph is analysed. The characterisation step takes under 10 seconds even for images with more than 600 dendrite cores.

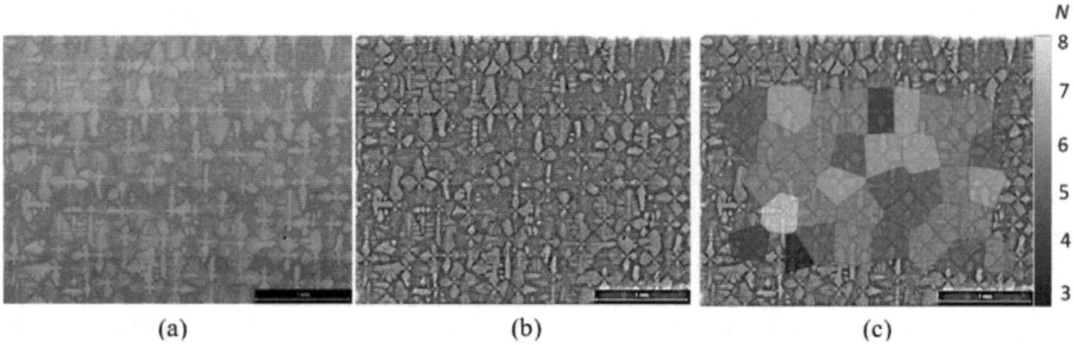

Figure 2. Diagram showing the complete process optimisation route of DenMap from a) CMSX-4® micrograph, b) dendritic core mapping and c) characterised micrograph demonstrating a combination of irregular geometries triangular (N_3), square (N_4), pentagonal (N_5), heptagonal (N_7) and octagonal (N_8).

3. Results

It is established that in order to gain sufficient stereological data about a single crystal alloy, the entire sample surface must be analysed. In this work, dendritic mapping (DenMap) and single crystal characterisation (SLPS) are performed on: CMSX-4® and CMSX-10® superalloys provided by Rolls-Royce Plc, on binary alloys from the literature: Pb-5.8 wt% Sb [10], Pb-10 wt% Sn [7], Al-4.1 wt% Cu [8] and on a SCN-0.24 wt% camphor cellular alloy cast under microgravity conditions [11].

The local shape factor, K, for all alloys is plotted against N in figure 3. The K_{SLPS} ideal line in figure 3 is derived from regular polygon shapes and indicates the optimum packing arrangement. All alloys demonstrate a tendency towards K_{SLPS} ideal with increasing N and almost an identical relationship which follows closely the optimal values. As observed, the N_3 and N_4 shapes for all alloys have much higher K values than the ideal.

The local PDAS range, $\lambda min / \lambda max$, is another dimensionless quantity by which dendritic arrays can be characterised, as shown in figure 4. The PDAS range is sorted by packing type and plotted with frequency, where the CMSX-4® obtains a maximum PDAS range of 3.8. The cellular SCN-24%wt camphor alloy cast in microgravity achieves a PDAS range below 2. An interesting observation is that the refinement of an alloy i.e. the reduction in the bulk PDAS range, is accompanied by an increase in the frequency of the N_5 shapes.

4. Discussion

The dimensionless local polygon shape factor (K_{SLPS}) provides a measure of the irregularity of packing around each individual dendrite, as in equation (1). Any value for K, larger that the optimum, designates a dendrite that on average has a smaller A_{TNN} and a higher $\bar{\lambda}$, indicating non-uniform microsegregation. This indicates zones of overgrowth, tip splitting or branching [12, 13]. The refinement of K with increasing N is expected, as larger N shapes constrict the dendrite arrangement and reduce the range of available irregular geometries. The CMSX-4® alloy on average demonstrates the largest variation in possible irregular shapes, closely followed by the CMSX-10® alloy. Both alloys have a complicated chemical composition with more than 6 alloying elements which leads to a much larger preferential microsegregation compared to any binary alloy. Microsegregation is known to cause convection, dendrite misorientation and other thermo-solutal effects. The SCN alloy solidified under near equilibrium conditions in microgravity where the convection effects were eliminated. It demonstrates a large reduction in the highly constricted N_7 and N_8 shapes, figure 4(a), and the average packing for each N was the closest out of all alloys studied to the optimal shape factor, K (figure 3).

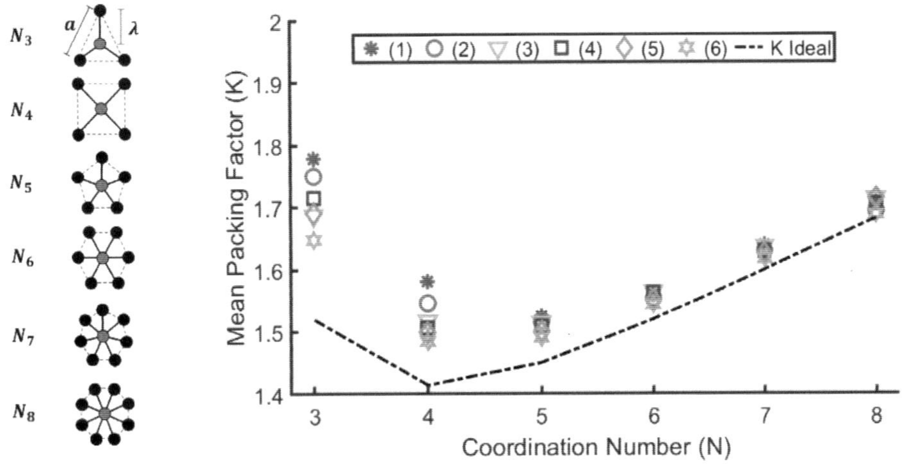

Figure 3. Six single crystal microstructures are investigated using SLPS: (1) CMSX-4®, (2) CMSX-10®, (3) Al – 4.1wt%Cu, (4) Pb – 10wt%Sn, (5) Pb – 5.8wt%Sb, (6) SCN – 0.24wt%Camphor. The mean packing factor for the seven main types of packing within each single crystal microstructure is plotted against the ideal packing factor, K_{SLPS}, for each regular polygon (dashed line).

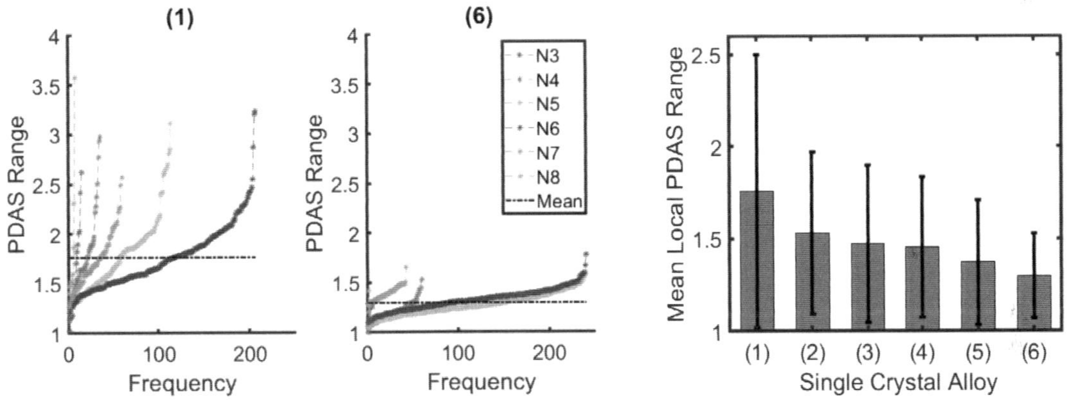

(a) (b) (c)

Figure 4. (a) Local PDAS range all determined packing patterns for (a) CMSX-4® and (b) SCN – 0.24wt%Camphor, (c) mean local PDAS range (MLPR) and variation for (1) CMSX-4®, (2) CMSX-10®, (3) Al – 4.1wt%Cu, (4) Pb – 10wt%Sn, (5) Pb – 5.8wt%Sb, (6) SCN – 0.24wt%Camphor. The dendritic CMSX-4® alloy possesses the largest mean range and variation whilst the SCN-camphor alloy cast under microgravity conditions the smallest.

A local PDAS range equal to 1 indicates the ultimate lower packing limit where all neighbouring dendrites or cells are equally spaced from the central one. Such a ratio is only achieved in a regular hexagonal arrangement (figure 3). Unsurprisingly, a single crystal alloy which possess a PDAS range for all determined *N* closer to that ultimate limit must possess a bulk microstructure that is more refined. A refinement in the single crystal alloy indicates a more homogenous distribution of interdendritic and dendritic material [14]. Therefore, figure 3 and figure 4 can be utilised to indicate how close/far an array is to the ideal hexagonal packing formation. Consequently, the CMSX-4® sample with the largest PDAS range and the highest average K value would possess the largest amount of non-uniform microsegregation, visualised as a highly non-homogenous microstructure.

Naturally, dendrites or cells are driven by thermo-physical conditions and tend to arrange in the most stable configuration as possible. In single crystal alloys ideal close packing is achieved when all local PDAS are equivalent i.e. hexagonally packed. Consequently, alloy refinement is accompanied by a decrease in unrefined N_3, N_4, N_7 and N_8 packing types and a subsequent increase in close packed N_5 and N_6 arrangements (figure 4). It is theorised that in a refined array, N_6 shapes are only formed from N_5 via translation or tip splitting [15] rather than overgrowth; the former phenomena occurring much slower than the latter. It is suggested that as convection is reduced, the bulk hexagonality of the array is increased and PDAS range continue to refine as shown in figure 4(b).

5. Conclusion

A novel dendritic mapping (DenMap) and characterisation digital tool (Shape-Limited Primary Spacing) is applied to a variety of single crystal microstructures. Several new dimensionless parameters are developed to identify new geometric relationships between any single crystal alloy cast under any condition. The results demonstrate that dendritic packing can be utilised to normalise statistical data to provide meaningful and accurate comparisons between all types of single crystal microstructure. Now, with a new standardising method for single crystal characterisation, large quantities of data can be processed, enabling the development of sophisticated process-property models for alloy optimisation.

Acknowledgements

We gratefully acknowledge the financial support from the Centre for Doctoral Training in Innovative Metal Processing (IMPaCT) funded by the UK Engineering and Physical Sciences Research Council (EPRSC) and Rolls-Royce Plc for providing the CMSX-4® and CMSX-10® samples.

6. References

[1] Mason JT, Verhoeven JD and Trivedi R 1982 *J. Cryst. Growth* **59** 516-24
[2] Somboonsuk K, Mason JT and Trivedi R 1984 *MMTA* **15** 967-75
[3] Tschopp MA, Miller JD, Oppedal AL and Solanki KN 2014 *Metall. Mater. Trans. A* **45** 426-37
[4] Nenchev B, Strickland J, Tassenberg K, Perry SS, Gill S, Dong HB 2020 *J. Imaging* ***6*** *19*
[5] Strickland J, Nenchev B, Dong HB 2020 *MCWASP XV (Accepted)*
[6] Trivedi R and Somboonsuk K 1985 *Acta Metall.* **33** 1061-8
[7] Trivedi R, Mazumder P and Tewari S 2002 *Metall. Mater. Trans. A* **33** 3763-75
[8] Tewari SN, Weng Y-H, Ding G and Trivedi R 2002 *Metall. Mater. Trans. A* **33** 1229-43
[9] McCartney DG and Hunt JD 1981 *Acta Metallurgica* **29** 1851-63
[10] Hui J, Tiwari R, Wu X, Tewari S and Trivedi R 2002 *Metall. Mater. Trans. A* **33** 3499-510
[11] Bergeon N, Tourret D, Chen L, Debierre J-M, Guérin R, Ramirez A, Trivedi R 2013 *Phys. Rev. Lett.* **110** 226102
[12] Han SH and Trivedi R 1994 *Acta Metall. Mater.* **42** 25-41
[13] Gandin C-A, Rappaz M and Tintillier R 1993 *Metall. Trans. A* **24** 467-79
[14] Wilson B, Cutler E and Fuchs G 2008 *Mat. Sci. Eng. A* **479** 356-64
[15] Weidong H, Xingguo G and Yaohe Z 1993 *J. Cryst. Growth* **134** 105-15

Dendritic Packing and Porosity Formation within Single Crystal CMSX-4® Ni-Base Superalloy

S Perry[1*], J Strickland[1], B Nenchev[1], K Tassenberg[1], HB Dong[1]

[1]School of Engineering, University of Leicester, Leicester, LE1 7RH, UK

sjp78@leicester.ac.uk

Abstract. Specialised imaging techniques are instrumental for extracting the wealth of information locked within as-cast microstructures. Recently, a novel characterisation tool (Shape-Limited Primary Spacing) was developed for determining primary dendrite arm spacing (PDAS), dendritic packing pattern and porosity formation within single crystal microstructures. In this work, the algorithm is applied to a single crystal CMSX-4® Ni-base superalloy to investigate the relationship between dendritic packing pattern and condensation porosity. It is concluded that porosity formation is reduced when regular organised dendritic packing patterns form. This work elucidates new unknown relationships between packing and porosity and may help to develop improved process-property models for defect prediction.

1. Introduction

Single crystal casting is a critical metal manufacturing technique used throughout the aerospace industry. High strength turbine blades produced via single crystal casting possess significantly higher creep resistance in comparison to directionally-solidified and equiaxed counterparts [1]. A continual goal for materials scientists is to increase jet engine performance and efficiency. Single crystal Ni-base super alloys contain significant levels of microsegregation; a result of their complex chemistry. The primary dendrite arm spacing (PDAS) is a key length scale used for characterisation of microsegregation between dendritic and the interdendritic regions. A large PDAS corresponds to increased microsegregation which results in the formation of low melting point secondary phase eutectics, as well as incoherent precipitates and pores in the interdendritic region. It is these secondary phases and defects which determine the properties of the as-cast material and its high temperature performance [2].

Porosity formation is an unavoidable phenomenon in any cast microstructure. It is a result of thermal strains and shrinkage which occurs due to lack of liquid feeding at the last to solidify regions [3]. In single crystal samples, porosity is localised in the interdendritic regions of the array [4]. This microporosity acts as crack initiation sites and is severely detrimental to material mechanical properties. Particularly important to turbine blade performance is the detrimental impact of porosity on fatigue life [5]. Classically, porosity characterisation constitutes identifying the pore type, be it a condensation, diffusion or deformation related formation. This involves careful microscale analysis to quantify the surface, shape and pore roughness to draw conclusions about the formation mechanisms. Pore formation is dictated by the liquid feeding capability to the interdendritic region, therefore, it is intrinsically related to the arrangement of the local dendritic neighbourhood. Dendritic packing patterns are a result of the bulk solidification conditions. Consequently, porosity formation must also be closely related to global variations in solidification conditions.

In this work, a recently developed Shape-Limited Primary Spacing (SLPS) algorithm [6] is applied to a single crystal CMSX-4® microstructure to relate dendritic packing patterns and porosity formation. The main aim of this work is to prove porosity and dendritic packing patterns are related; thus, improving the understanding of porosity formation mechanisms. The end goal is to improve single crystal modelling capability through further understanding pore formation pathways. This will reduce scatter of as-cast/solutionised mechanical property data, thereby allowing turbine blade design service stresses to be raised.

2. Experimental procedure

A single crystal as-cast CMSX-4® bar of 9.4 mm diameter was sectioned using a metallographic precision cutter and mounted in Conducto-Mount. The samples were prepared for Scanning Electron Microscopy (SEM) through standard metallographic preparation and polished using 1um diamond suspension. The sample was imaged in backscattered electron (BSE) mode using a FEI 650 Quanta SEM facility at the University of Leicester (20 keV, 5.0 spot, 1.92mm HF, 10.1 mm WD) to visualise the microsegregation between the dendritic core and interdendritic region. The final image was comprised of a 24 $\times$ 24 grid of individual SEM images taken at 238 $\times$ magnification with 20 μs scan speed and encompassed the full horizontal plane to the test bar. The image was loaded into an automatic dendritic core detection software, DenMap, and the exact core and porosity pixel positions were located [7]. The dendritic packing and local PDAS were calculated using the novel Shape-Limited Primary Spacing (SLPS) algorithm [6] which is based on Voronoi area and regular polygon shapes (figure 1). The algorithm identifies the dendrites whose diffusion fields interact locally with a central dendrite. This method of true nearest neighbour (TNN) determination is applied to each dendrite within the single crystal array. The porosity area and location are indexed and overlapped on each local packing pattern (Voronoi shape) such that each pore belongs to one polygon. This allows local and bulk spacing and packing to be related to pore location.

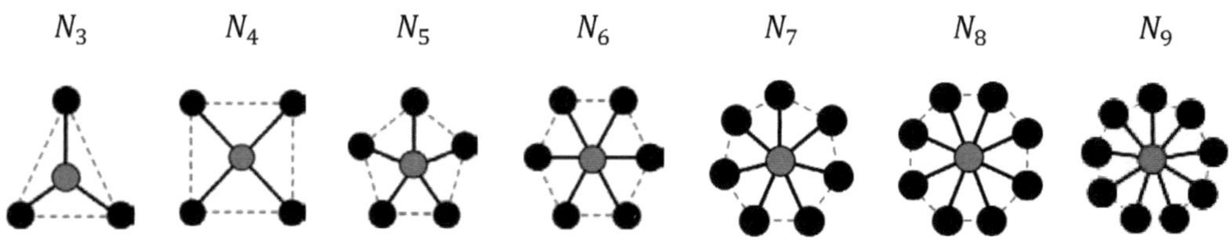

Figure 1. Dendritic packing patterns. Coordination number, N, indicates the true nearest neighbouring dendrites determined by the Shape-Limited Primary Spacing (SLPS) algorithm.

3. Results

The packing results are illustrated in figure 2(a), the scale bar indicates the type of dendritic packing pattern. The top right of figure 2(a) demonstrates refined connected hexagonal packing (N_6). The bottom left of figure 2(a) a combination of irregular geometry triangular (N_3), square (N_4), pentagonal (N_5), heptagonal (N_7), octagonal (N_8) and nonagonal (N_9) packing. Figure 2(b,) illustrates the detected pore size within figure 2(a). Pore areas of about $\sim$50 μm are the most frequently detected. Table 1 indicates the N_6 region (section 1) obtains a more refined PDAS, six times smaller pore frequency and a smaller average and maximum pore size than the irregular packed region (section 2).

The bulk of the pores are congregated on the left-hand side of the sample (figure 3(a)). There is a region of increased pore frequency shown in green in the bottom half of figure 3(a). The upper right-hand side demonstrates almost no pores, which also corresponds to the hexagonally packed, N_6, region in figure 2(a). N_6 packing over the entire sample demonstrates the smallest pore frequency per shape out of all determined coordination numbers, N. A pore area distribution plot is illustrated in figure 3(c). Similarly, to figure 3(a) the largest pore sizes are congregated in a band across the bottom half of the sample in figure 3(c). The N with the largest pores is determined as the triangularly packed

N_3 dendrites. The smallest pore area per shape area is determined as N_8 which is closely followed by N_6.

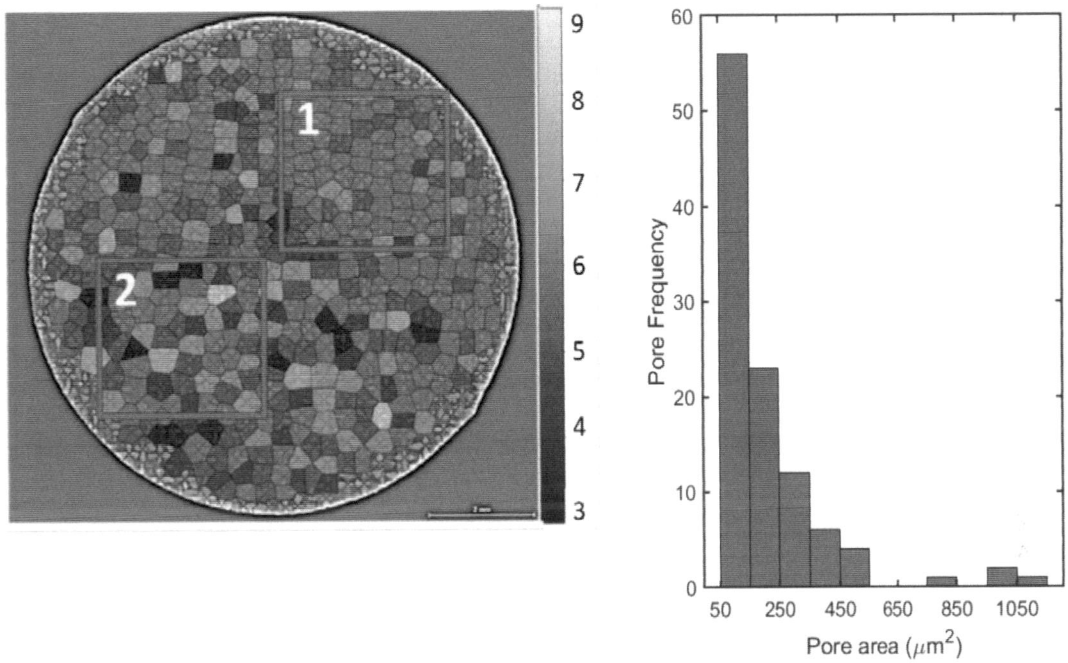

(a) (b)

Figure 2. (a) Shape-Limited Primary Spacing (SLPS) Voronoi packing map. Key data for the highlighted sections (1 & 2) are illustrated in Table 1. (b) Pore frequency vs pore area for the entire sample in figure 2(a).

Table 1. Mean PDAS and pore data for section 1 & 2 from figure 2(a).

Region	Packing	Mean PDAS (μm)	Pore Frequency	Average Pore Size (μm^2)	Maximum Pore Size (μm^2)
1	Hexagonal	311.91	17	65.11	269.09
2	Irregular	367.56	102	79.28	981.19

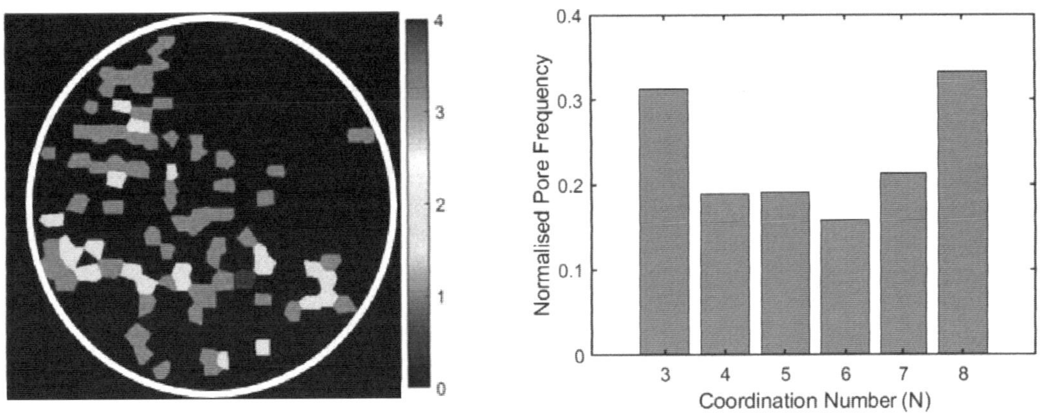

(a) (b)

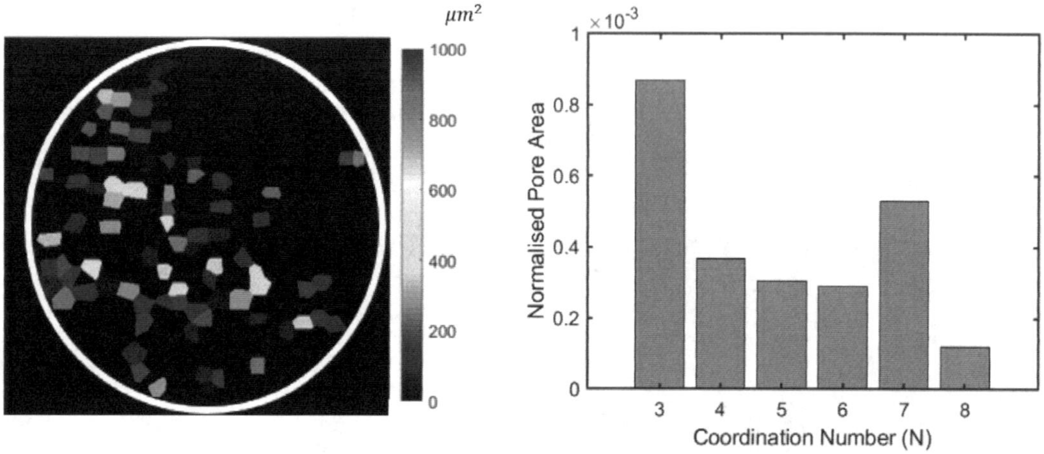

(c) (d)

Figure 3. (a) Pore frequency distribution plot. The red region indicates a Voronoi shape with 4 pores within its area. The plot provides no measure of pore size. (b) Normalised pore frequency by coordination number, N. (c) Pore area distribution plot. The scale is measured in μm^2. The red areas indicate regions with largest pore area. (d) Normalised pore area by N.

4. Discussion

A variety of dendritic packing has been determined across the entire microstructure (figure 2(a)). There exists a clear hexagonally packed, N_6, region in the top right of the micrograph, and an irregular packed region in the bottom left. Interestingly, the majority of the microporosity resides in the irregularly packed regions, illustrated figure 3(a,c). In fact, in the irregularly packed region in section 2, there exists six times the microporosity as the regular packed hexagonal region (table 1). Porosity is known to be intrinsically linked to fluid flow during solidification [8]. Therefore, the organised arrangement of dendrites in the hexagonal configuration is likely to allow sufficient liquid feed to the last to solidify regions which in turn reduces the propensity for microporosity formation.

Porosity formation within the N_3 packing was the second most frequent observed within the sample (figure 3(b)) and it obtained the largest pore area per packing size (figure 3(d)). Interestingly, N_3 dendrites surround the higher order N_8 and N_7 packing patterns which also witness the second and third largest pore frequencies, respectively (figure 3(b)). These packing patterns are all congregated together on one side of the sample. It follows that porosity mainly forms between the N_3, N_7 and N_8 dendrites. The N_7 and N_8 configurations have the largest associated areas on the micrograph, clearly seen in figure 2(a). These arrangements with large PDAS indicate areas rich in interdendritic partitioning elements such as Ni, Ti, Ta, and Al [9]. Consequently, the close packed areas with small PDAS indicate regions rich in dendritic core elements such as Co, W and Re [10]. It follows, that the largest and most frequent porosity forms between the N_3 and N_7 and N_8 dendrites. Therefore, the most detrimental porosity to material mechanical properties forms when large interdendritic rich regions are neighboured by correspondingly smaller core element rich regions. To fill shrinkage voids, melt must flow downwards between neighbouring dendrites [11]. The irregular packed shapes inhibit the ability of the liquid to flow and fill these voids. It follows, that the more irregular the dendritic packing arrangement the more liquid feed is restricted and the higher the propensity for porosity formation.

The temperature profile across a sample during solidification is uneven. A strong radial thermal gradient component exists due to the nature of the directional solidification process [12]. Therefore, the solidification rate across a sample section varies, with one side of the sample solidifying more quickly than the other. Interestingly, a band of pores exists across the microstructure which finds itself between the regular and irregular packed regions (figure 3(a)). This porosity band witnesses not only the highest frequency of pores but also the largest (figure 3c). Therefore, microporosity formation may be intrinsically linked to differential rates of thermal contraction due to solidification strain. The disparity in freezing point which exists between the separate regions may induce a tensile stress in the sample which manifests in these oversized pores. Further investigation into these zones is needed to establish the exact relationship between thermal contraction, shrinkage and microsegregation. However, DenMap, SLPS and the shown pore-packing analysis are capable of locating and characterising these crucial zones.

5. Conclusions

A novel packing and porosity algorithm, Shape-Limited Primary Spacing (SLPS) is applied to study a CMSX-4® single crystal Ni-base superalloy. The new technique has elucidated unknown relationships between dendritic packing patterns and the corresponding porosity formation. For successful modelling of defect prediction, a full understanding of the influence of solidification conditions on the propensity for defect formation is required. SLPS enables rapid characterisation of local PDAS, packing patterns and defects, within single crystal alloys. Now, for the first-time single crystals can be characterised in a way that is repeatable, comparable, and accurate. Therefore, enabling development of sophisticated defect prediction models, machine learning algorithms and eventually A.I.

Acknowledgements

We gratefully acknowledge the financial support from the Centre for Doctoral Training in Innovative Metal Processing (IMPaCT) funded by the UK Engineering and Physical Sciences Research Council (EPRSC).

6. References

[1] Toloraiya VN, Orekhov NG and Kablov EN 2002 *Met. Sci. Heat. Treat.* **44** 279-83
[2] Lamm M and Singer R 2007 *Metall. Mater. Trans.* A **38** 1177–83
[3] Carlson KD and Beckermann C 2009 *Metall. Mater. Trans.* A **40** 163-75
[4] Rappaz M, Jacot A and Boettinger WJ 2003 *Metall. Mater. Trans.* A **34** 467-79
[5] Ruttert B, Meid C, Mujica Roncery L, Lopez-Galilea I, Bartsch M and Theisen W 2018 *Scr. Mater.* **155** 139-43
[6] Strickland J, Nenchev B and Dong HB 2020 *MCWASP XV* (Accepted)
[7] Nenchev B, Strickland J, Tassenberg K, Perry S, Gill S, and Dong HB 2020 *J. Imaging* **6** 19
[8] Rappaz M, Drezet J-M and Gremaud M 1999 *Metall. Mater. Trans.* A **30** 449-55
[9] Kearsey RM, Beddoes JC, Jones P and Au P 2004 *Intermetallics.* **12** 903-10
[10] Trivedi R and Somboonsuk K 1985 *Acta. Metall. Sin.* **33** 1061-8
[11] Dantzig J and Rappaz M 2017 *Solidification* (Switzerland: EPFL Press) 2nd 548-9
[12] Ridolfi MR, Tassa O and de Rosa G 2017 *Mater. Sci. Forum.* **879** 1582-7

A Novel Approach for Determination of Hardening Parameters of an Aluminum Alloy under Cyclic Loading with High Amplitudes

Eray Arslan[1,]*, Milan Zigo[2] and Gerhard Kepplinger[2]

[1]Institute of Mechanics and Mechatronics, TU Wien, Vienna, Austria
[2]MAGNA STEYR, Graz, Austria

*eray.arslan@tuwien.ac.at

Abstract. An optimization procedure for the determination of combined nonlinear isotropic/kinematic hardening parameters for strain rate-independent cyclic plasticity has been investigated. With this procedure, it is offered that only nine material parameters, including Young's modulus, yield stress, Poisson's ratio, two isotropic hardening parameters, and four kinematic hardening parameters, will be sufficient to perform cyclic plasticity analysis with high accuracy in well-known FE programs (e.g. ABAQUS) even for separate strain range amplitudes. The procedure has been performed to calculate such parameters for aluminum alloy EN AW-6082, which is frequently used in vehicle bodies and simulation results are compared with experiment results. The determined parameters lead highly accurate simulation results in FE routines.

1. Introduction

High strength aluminum alloys are used in automotive components frequently [1]. In the most of applications, such components are subjected to cyclic loads, some of which are with high strain amplitudes. Durability of these components may be influenced under different load cases [2]. They may be subjected to initially small number (3-12) of load cycles but with very high amplitude. Then it is required to evaluate the permanent deformation and the crack (breakage of the component) after repetitive high plasticization. In another case, structures may work under further large number of load cycles with small amplitudes after a high plasticization case at the beginning. In this case, by consideration of the history for their plastic deformations, hardening behavior plays a significant role for the further fatigue life calculations and evaluation of permanent deformation. In any case, hardening parameters of the materials should be determined accurately and taken into account in the further durability analysis, which is performed in FE simulations.

In the literature, there exists several investigations presenting procedures to obtain material parameters to be used in the cyclic plasticity applications [3,4]. However, in the most of such investigations [3,5] parameters may be valid only for one strain range amplitude cyclic plasticity applications. These parameters may lead very accurate simulation results for certain strain amplitudes but not for different strain ranges. On the other hand, in some other studies in the literature [4], the material parameters and the theory behind to derive those parameters may not be compatible to well-known FE software (e.g. ABAQUS [6]) and this is impractical in the most of engineering applications. Importance of this issue has been already detected by Badnava et al. [7] and then a generic algorithm has been developed to determine hardening parameters for stainless steel. Furthermore, general classical mathematical formulation in the literature may be useful for ideal test data (e.g. such as in [8]) however, imperfect [9] or uncharacteristic behavior of such materials require a particular optimization routine to determine accurate material properties [10]. Another issue can be shown as the high number of parameters [11] to use in the evaluation of hysteresis behavior. Decreasing the number of the parameters will be practical in the engineering applications.

Aim of the present study is to develop a model which is capable to describe low cycle fatigue behavior for practical usage on real metallic components (e.g. EN AW-6082). In this model, a novel optimization procedure has been performed for the determination of the combined nonlinear isotropic/kinematic hardening parameters [8,12] for strain rate-independent cyclic plasticity. Only nine material parameters (Young's modulus, yield stress, Poisson's ratio, two isotropic hardening parameters, and four kinematic hardening parameters) lead to perform cyclic plasticity simulations in FE programs (e.g. ABAQUS) accurately even for separate strain amplitude ranges.

In the study, firstly hysteresis curves of aluminum alloy specimens have been determined with a uniaxial tension and compression test. Test data has been used to calculate initial isotropic and kinematic hardening parameters by using classical mathematical formulation in the literature. These parameters initiate optimization routines to determine more accurate parameters in FE simulation. Final simulation results are compared with the test data in a hysteresis curve for aluminum alloy EN AW-6082.

2. Experiment

The aim of the experiment is to determine hysteresis curve of aluminum alloy specimens under uniaxial tension and compression loads at room temperature. The experiment has been performed for three different strain amplitudes which are ±1 %, ±3%, and ±5 % under strain controlled test conditions. Cyclic stress-strain curves are established for each specimen by incremental-step method [12] in which the strain amplitude is held constant until its stabilization then the procedure is repeated for the other strain range amplitudes on the same samples. To do so three cycles have been completed at each strain amplitude in a constant quasi-static strain rate for a single specimen. Then the test data is prepared to be used in the following mathematical model effectively.

3. Mathematical Model

Combined nonlinear isotropic/kinematic hardening model [8,12,13] has been calibrated to determine deformation behavior of the structure subjected to cyclic loading in quasi-static loading rate. In order to calculate hardening parameters initially Young's modulus and initial yield stress at zero plastic strain should be determined from the test data by consideration not only the loading but also unloading regions. Poisson's ratio is taken as $\nu = 0.33$ for aluminum alloy. Formulation for the isotropic and kinematic hardening models has been followed separately to determine parameters which lead for the initial values of the optimization routines.

In the isotropic hardening model, the change of the size of the yield stress σ^0 with equivalent plastic strain ε^p has been defined by

$$\sigma^0 = \sigma_y + Q_\infty \left(1 - e^{-b_{iso}\varepsilon^p}\right) \tag{1}$$

where σ_y is initial yield stress at zero plastic strain. This equation is used to calibrate initial isotropic hardening parameters Q_∞ and b_{iso} [12]. In this formulation, the test data for one strain range amplitude but each cycle until stabilization are taken into consideration. During the calibration least squared regression has been used.

According to classical kinematic hardening model, Frederick-Armstrong formulation is used to define relationship between stress difference (between maximum tension and compression stresses $\Delta\sigma$) and the corresponding plastic strain differences ($\Delta\varepsilon^p$) at each strain amplitude range [8, 14,15]

$$\Delta\sigma = \sum_{i=1}^{2} \frac{2C_i}{\gamma_i} \tanh\left(\frac{\gamma_i}{2}\Delta\varepsilon^p\right) + 2\sigma_y \tag{2}$$

where C_i and γ_i are kinematic hardening parameters. The formulation has been performed with the consideration of two back stresses ($i:1-2$). It is noted that only one back stress is inaccurate for the fitting but two back stresses are optimum for the accuracy and the number of parameters. During the calibration least squared regression has been used.

The hardening parameters, which have been calculated by the classical mathematical formulation above, are used for the initial estimation in an optimization routine for the determination of more accurate parameters. To do so, optimization routines have been used with least squares error calculations. To accelerate the routines, effect of each particular parameter on the hysteresis curve has been taken into consideration. Parametrization loops in FE model have been performed to determine the most accurate hardening parameters. The results of the FE simulations have been evaluated and compared with the test data, then an error between those results are determined. The optimization routine has been performed to reduce this error. Effects of all material parameters have been evaluated and correspondingly optimum hardening parameters have been determined. Even though the initial parameters which are calculated with the classical mathematical formulation above leads around 40% error, optimized parameters cause to decrease of the corresponding error to 4%.

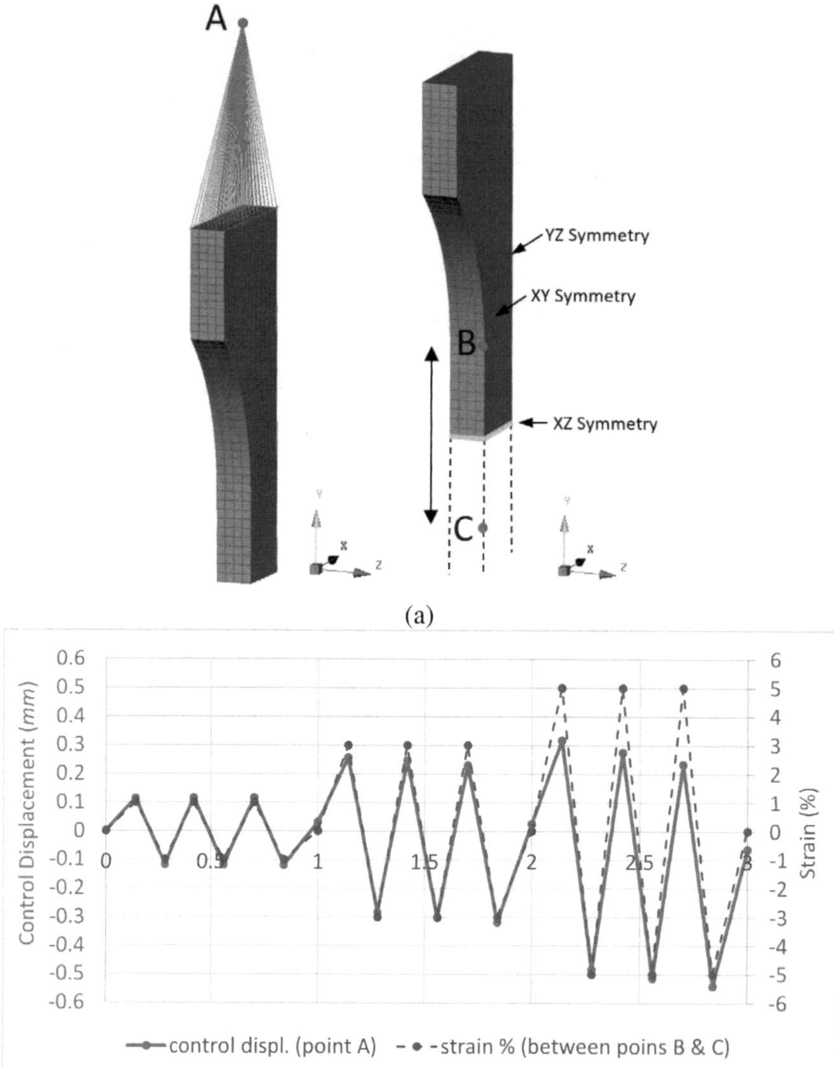

(b)

Figure 1. (a) FE Model, (b) Input displacement at control point A (red line) and axial strain measured between points B and C (blue dashed line)

Material parameters are used in an FE model (parallel to optimization routine) to simulate hysteresis behavior under same loading conditions with the experiment. Three plane symmetries in specimen have been considered as one can see in Figure 1(a). In this figure, point A denotes the control point of the model to apply input displacement, and at each step axial strains between points B and C are measured. To fit the measured strain amplitude with the test data, the input displacement amplitude is rearranged. Hence, strain control conditions have been provided in the FE simulation. Corresponding input displacement amplitude (at point A) and measured strains (between points B and C) are plotted in Figure 1(b).

4. Numerical Results

Young's modulus *E*, yielding stress σ_y, isotropic hardening parameters (Q_∞ and b_{iso}) and kinematic hardening parameters for two backstresses case (C_i and γ_i where $i = 1, 2$) for aluminum EN AW-6082 are calculated first by the classical mathematical formulation and then optimized by using the methodology in the previous sections. Initially calculated material parameters of the material by using classical mathematical formulation have been presented in Table 1.

Table 1. Calculated initial material parameters of aluminum alloy EN AW-6082 for cyclic plasticity

E (GPa)	σ_y (MPa)	ν	Q_∞	b_{iso}	C_1	γ_1	C_2	γ_2
66.5	227.42	0.33	27.7	88.5	11436.4	167.0	3738.2	110.4

The parameters in this table are used directly in FE simulation to determine initial hysteresis behavior of a test specimen for different strain amplitudes (±1 %, ±3%, and ±5%). Result of this simulation and its comparison with the corresponding test result are presented in Figure 2. As one can see in this comparison, the simulation results do not match with the test data, perfectly. Average error between results is around 39 %.

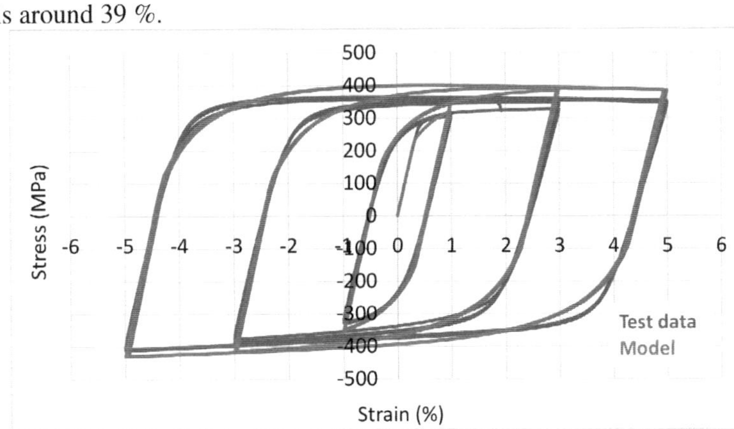

Figure 2. Comparison of hysteresis curves of test data (blue line) and results of suggested model (red line) for EN AW-6082 with initially determined parameters.

However, optimized material parameters, as given in Table 2, represent relatively better agreement between simulation results and test data as seen in Figure 3. It should be emphasized that average error between simulation and test results for all strain ranges are not higher than 4.2 % in this case.

Table 2. Optimized material parameters of aluminum alloy EN AW-6082 for cyclic plasticity

E (GPa)	σ_y (MPa)	ν	Q_∞	b_{iso}	C_1	γ_1	C_2	γ_2
66.5	245	0.33	39.54	5.03	14363.6	595.9	6031.5	103.8

5. Conclusion

Nonlinear isotropic and kinematic hardening parameters have been identified by using classical hardening models to determine cyclic plasticity behavior of an aluminum alloy. Then the parameters have been optimized by using a realistic FE simulation in ABAQUS. Finally, simulation results show good agreements with the experiment even though only nine material parameters have been used in the simulations. It should be emphasized that the determined parameters represent accurate responses for both low and high strain amplitudes. It is also noted that the procedure may be applied for any imperfect test data or materials which represent uncharacteristic hysteresis behavior to obtain realistic and accurate FE simulations.

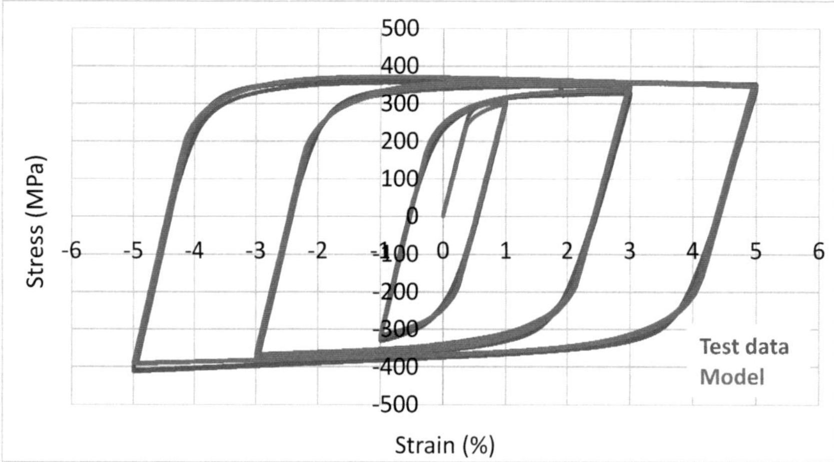

Figure 3. Comparison of hysteresis curves of test data (blue line) and results of suggested model (red line) for EN AW-6082 with optimized parameters.

Acknowledgements

This work was supported by Austrian Research Promotion Agency (FFG) and MAGNA STEYR with the funding project entitled 'Next Generation of Virtual Product Development (ViProD 2.0)' with the project number 855745.

6. References

[1] Hägele N and Sonsino C M 2014 *Int. J. Fatigue* **69**, 63
[2] Mastinu G and Plöchl M 2014 *Road and Off-Road Vehicle System Dynamics Handbook*, (New York: Taylor and Francis)
[3] Mahmoudi A H, Pezeshki-Najafabadi S M and Badnava H 2011 *Comp. Mater. Sci.* **50** 1114
[4] Khademi E, Majzoobi G H, Bonora N and Gentile D 2015 *J. Strain Analy. for Eng. Des.* **50** 314
[5] Ashraf Q J, Prasad Reddy G V, Sandhya R, Laha K and Harmain G A 2017 *Fatigue Fract. Eng. Mater. Struct.* **40** 1
[6] ABAQUS Analysis User's Manual v2016
[7] Badnava H, Pezeshki S M, Fallah Nejad K and Farhoudi H R 2012 *J. Mech. Sci. Tech.* **26** 3067
[8] Budahazy V and Dunai L 2013 *Periodica Polytech.* **57** 139
[9] Arslan E, Mack W, Zigo M and Kepplinger G 2017 *Proc. Appl. Math. Mech.* **17** 385
[10] Goyal S, Gupta S K, Sivaprasad S, Tarafder S, Bhasin V, Vaze K K and Ghosha A K 2013 *Procedia Eng.* **55** 136
[11] Kang G, Ohno N and Nebu A 2003 *Int. J. Plast.* **19** 1801
[12] Nip KH, Gardner L, Davies C M, and Elghazouli A Y 2010 *J. Const. Steel Res.* **66** 96
[13] Cao J, Lee W, Cheng H S, Seniw M, Wang H P and Chung K 2009 *Int. J. Plast.* 25 942
[14] Hales R, Holdsworth S R, O'Donnell M P, Perrin I J and Skelton R P 2002 *Mater. at High Temp.* **19** 165
[15] Chaboche J L 2008 *Int. J. Plast.* **24** 1642

IOP Conf. Series: Materials Science and Engineering **947** (2020) 012010 doi:10.1088/1757-899X/947/1/012010

Mechanical and Thermal Characterization of Short Salago Fiber (*Wikstroemia spp.*) Reinforced Epoxy Resin Composites

J D Lopena[1] and J C Millare[1]

[1]School of Chemical, Biological and Materials Engineering and Sciences, Mapúa University, Muralla Street, Intramuros, Manila, 1002, Philippines

jerome_lopena@yahoo.com; jcmillare@mapua.edu.ph

Abstract. Natural fibers are used as alternative reinforcements in polymer composites due to their biodegradability, low density and cost. Applications of natural fiber reinforced polymers include aerospace industry, automotive parts and marine vessels. In the Philippines, one source of natural fiber is salago, which is a type of shrub with tough bark used to produce high quality paper products. In this research work, short salago fibers were reinforced in epoxy resin in random orientation. The fibers were treated with 5 wt.% NaOH solution. The fiber loading considered in the study are 5 wt.%, 10 wt.%, 15 wt.% and 20 wt.%. Mechanical properties of neat epoxy and salago fiber composites were characterized through tensile, flexural and impact tests. Thermogravimetric analysis was used to evaluate the thermal stability of the specimens. Results for 20 wt. % fiber loading show significant improvements on tensile modulus (83.5 %) and impact strength (26.4 %) as compared to neat epoxy. However, decrease on tensile and flexural strengths were observed as the fiber content increases. Nonetheless, noticeable improvement on thermal stability was obtained for the composite with 10 wt.% fiber content.

1. Introduction

In recent years, the applications of natural fiber reinforced composites (NFRCs) in industries have increased drastically due to the availability of natural fibers at lower costs as compared to synthetic materials [1]. Some applications of natural fibers include automotive parts, aerospace components, electronics cases, sports equipment, packaging and textiles [2]. NFRCs have properties that are comparable to synthetic materials like mechanical strength, thermal properties and tribological characteristics [3]. Mechanical properties of these composites depend on various parameters such as interaction between fiber and matrix, aspect ratio, fiber loading and orientation, manufacturing techniques, etc. [4]. Chemical treatment such as alkaline and silane can also be applied to natural fibers to enhance adhesion with the matrix which can further improve the properties of composites [5]. The commonly used natural fibers in polymers are coir, sisal, jute, banana and flax [6]. However, due to the increase in demand for natural fibers, there is a need to explore other sources of fibers and study their possible use in different matrices. In the Philippines, one of the sources of natural fiber is salago. It is a type of shrub in which the extracted bast fiber can be used to produce high quality paper products. In a study, it was found out that salago fiber has cellulose content of 79.05 % and tensile strength of 1187 MPa [7]. Salago fiber has also been studied as filler material in high-density polyethylene where the properties of the composites were characterized [8, 9]. Apparently, only few research works about the use of salago fiber as reinforcement or filler in composites were conducted.

IOP Conf. Series: Materials Science and Engineering **947** (2020) 012010 doi:10.1088/1757-899X/947/1/012010

2. Methods

2.1 Materials

Salago barks were obtained from a paper factory in Cavite, Philippines. Both epoxy resin and hardener were purchased from J. Spencer Technologies Philippines Corporation in Muntinlupa, Philippines. Sodium hydroxide and acetic acid were procured from a local supplier.

2.2 Fiber Separation and Treatment

Salago barks were cut into 10 cm long and washed with tap water to remove dirt. The barks were then subjected to water retting for one week. The fibers were separated manually and air dried for two weeks. The fibers were then soaked in 5 % NaOH solution for 30 minutes using a liquor to fiber weight ratio of 20:1. In order to neutralize the effect of NaOH, the fibers were immersed in 2 % acetic acid solution for 1 hour [10]. These were subsequently rinsed three times in distilled water. The alkali treated fibers were sun dried for 3 days and oven dried at 70 °C for 4 hours to remove excess moisture. The salago barks and their separated fibers are shown in figure 1 and figure 2, respectively.

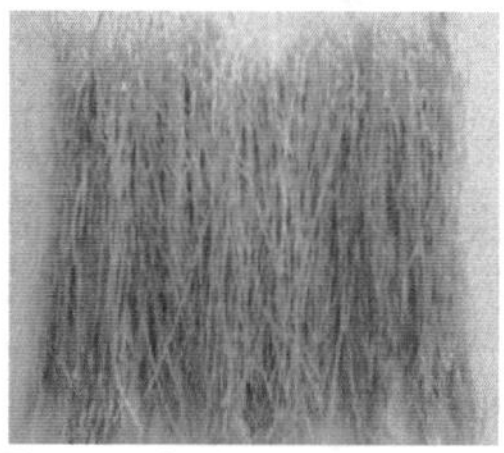

Figure 1. Salago barks. **Figure 2.** Separated fibers.

2.3 Specimen Preparation

An open molding process was used to fabricate the composites through silicone molds with shapes based on the dimensions specified in ASTM Standards D638 (Type I) and D790. The treated salago fibers were cut into 5 to 7 mm long and laid on the silicone molds in 2-D random orientation. The epoxy resin and hardener were mixed manually using the recommended weight ratio of 3:2 for 5 minutes before pouring the mixture into the silicone molds. A metal rod was used to distribute the resin to the fibers. The specimens were cured for 72 hours at room temperature. Varying fiber loading of 5 wt.%, 10 wt.%, 15 wt.% and 20 wt.% including neat epoxy were prepared. Figure 3 shows the salago fiber composites for tensile test.

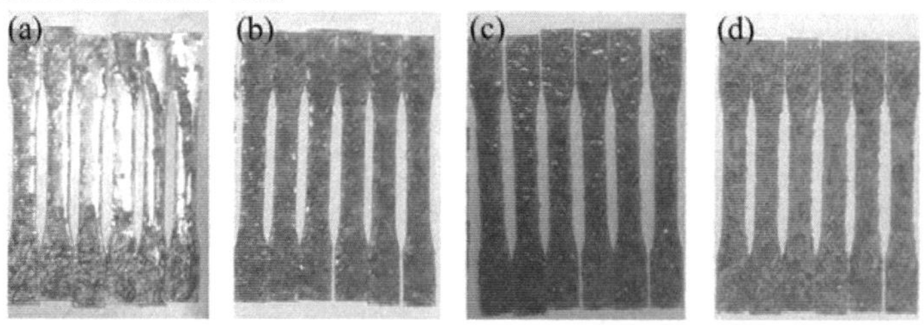

Figure 3. Salago fiber composites (a) 5%, (b) 10%, (c) 15% and (d) 20%.

2.4 Mechanical Testing

The specimens for mechanical testing were brought to the Standards and Testing Division of the Industrial Technology Development Institute of the Department of Science and Technology (DOST-ITDI) in Metro Manila, Philippines. Tensile test was performed based on ASTM Standard D638 using Instron UTM (Model:5585 H) with 5 mm/min speed of testing. On the other hand, flexural test was conducted based on ASTM Standard D790 using Shimadzu UTM (AGS-50kNXD) with a support span to depth ratio of 16:1. For Izod impact test, it was performed based on ASTM Standard D256 Method A using Zwick/Roell 5.5P Izod Pendulum Tester. Five trials were considered for each mechanical test. The average values and standard deviations were then calculated.

2.5 Thermogravimetric Analysis (TGA)

The specimens for TGA were brought to the Advanced Device and Materials Testing Laboratory (ADMATEL) of DOST-ITDI. TGA was performed using Perkin Elmer STA 6000. The temperature program used was from 30 °C to 950 °C with heating rate of 10 °C/min. The atmosphere considered was nitrogen at 20 mL/min. The initial weight of samples ranges from 10.4 mg to 11 mg.

3. Results and Discussion

3.1 Tensile Properties

Figure 4(a) shows that the tensile strength of the composites decreases as the fiber content increases. The average tensile strength obtained for neat epoxy is 49.10 MPa while 26.80 MPa for 20 wt.% fiber loading. The decrease in tensile strength is due to the considered length of salago fiber (5–7 mm) which could be less than the critical fiber length. This resulted to inefficient stress transfer between the fibers and matrix at higher tensile loads [4, 11]. On the other hand, the tensile modulus of epoxy (3.10 GPa) increases as the fiber content increases to 20 wt.% (5.69 GPa) as presented in figure 4(b). The improvement on the tensile modulus indicates good adhesion between the fibers and matrix since the composites with higher fiber contents resisted extension better than neat epoxy.

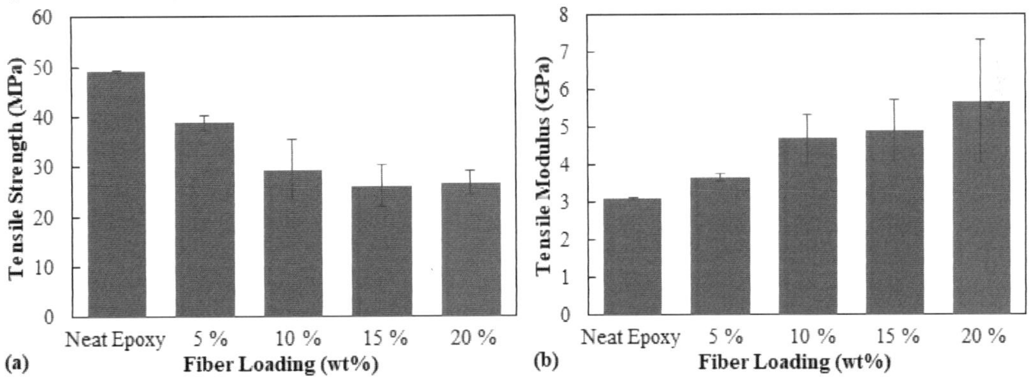

Figure 4. Tensile properties of salago fiber composites: (a) tensile strength and (b) tensile modulus.

3.2 Flexural Properties

Figure 5(a) shows that the addition of short salago fibers in epoxy resin decreases the flexural strength. The average flexural strength obtained for neat epoxy is 63.5 MPa while 32.3 MPa for 20 wt.% fiber loading. One factor that contributed to a lower strength is the 2-D random orientation of the short fibers. The composites became vulnerable to bending since the load direction is perpendicular with respect to the fiber length. For the flexural modulus of epoxy (2.11 GPa), there is a slight improvement at 10 wt.% (2.24 GPa) as shown in figure 5(b). This is because of the good fiber-matrix interaction. However, the decrease at 15 wt.% (1.54 GPa) is due to the effect caused by the 2-D fiber orientation.

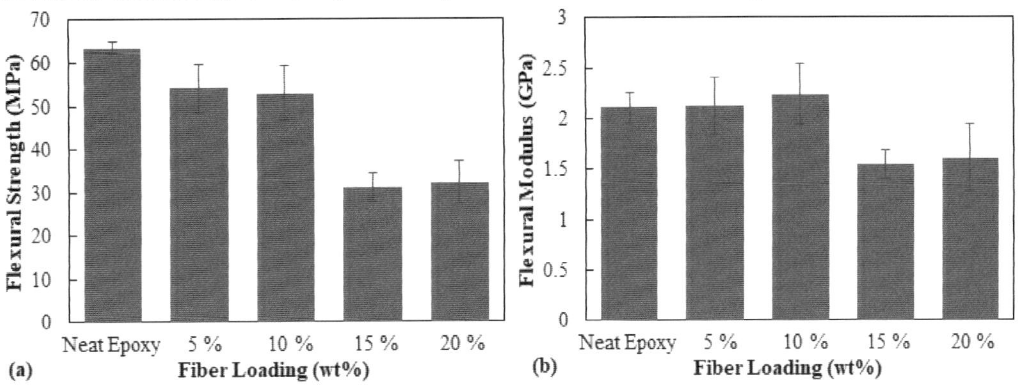

Figure 5. Flexural properties of salago fiber composites:(a) flexural strength and (b) flexural modulus.

3.3 Impact Strength

Figure 6 shows that the average Izod impact strength of neat epoxy (48.10 J/m) decreases at 5 wt.% fiber loading (31.7 J/m) but it increases at 10 wt.% (55.50 J/m) and 20 wt.% (60.8 J/m). The decrease in impact strength at 5 wt.% fiber content could be due to the tendency of the fibers to hinder deformation that resulted to a decrease in absorbed energy [12]. On the other hand, the improvement of Izod impact strength of the composites at 10 wt.%, 15 wt.% and 20 wt.% fiber loading is due to the good adhesion between the treated salago fibers and epoxy resin.

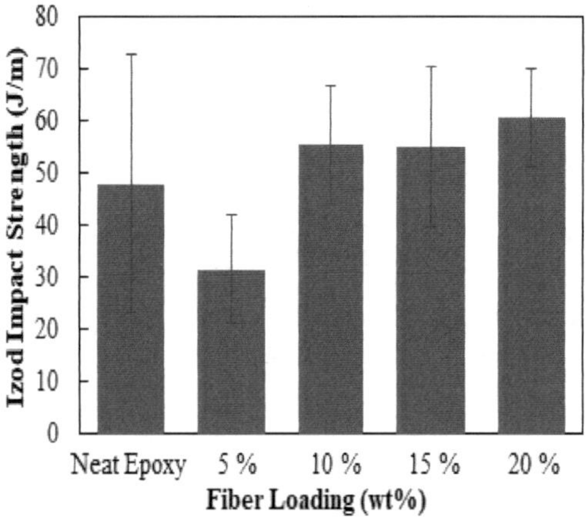

Figure 6. Impact strength of composites.

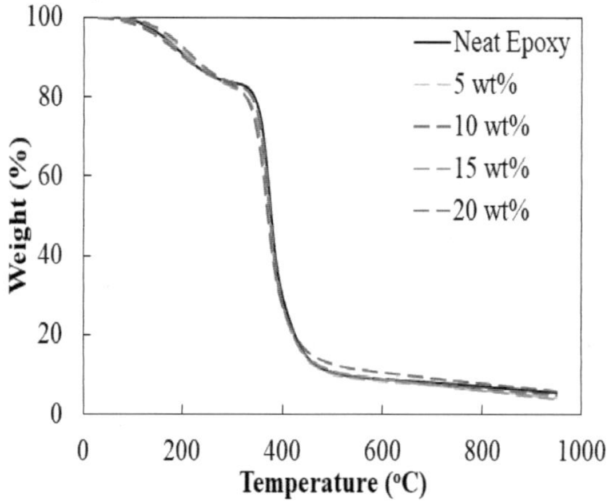

Figure 7. TGA curves of composites.

3.4 TGA Results

Results for TGA are presented in table 1 and figure 7. The initial weight loss observed in the samples is due to the removal of moisture and other present volatile components [13]. At the 1st degradation stage, there is an increase in peak temperature from 5 wt.% to 10 wt.% but it decreases as the fiber loading further increases from 15 wt.% to 20 wt.%. The 10 wt. % fiber loading has the highest initial peak temperature (207.69 ^{o}C) with a corresponding weight loss of 16.41 % as compared to the peak temperature obtained for neat epoxy (189.05 ^{o}C). This is because of the good interaction between the alkali treated fibers and epoxy resin which is also supported by the results obtained in the tensile modulus and impact strength of the composites. The addition of salago fibers at 10 wt.% in the epoxy matrix significantly increased the initial peak temperature by about 18.64 ^{o}C. This indicates improvement on the thermal stability of the composite. However, at the 2nd degradation stage, the peak temperature of epoxy (369.39 ^{o}C) decreases slightly as the salago fiber content increases from 5 wt.% (367.27 ^{o}C) to 20 wt% (364.23 ^{o}C). The corresponding weight loss at this stage is associated with the decomposition of cellulose and hemicellulose of the fibers [14]. At the final temperature of 950 ^{o}C, the amount of residue left ranges from 4.10 % to 5.95 %.

Table 1. TGA results of salago fiber reinforced epoxy resin composites.

Sample	1st stage		2nd stage		Residue at 950 ^{o}C (%)
	Peak temperature (^{o}C)	Weight loss (%)	Peak temperature (^{o}C)	Weight loss (%)	
Neat Epoxy	189.05	16.24	369.39	78.09	5.67
5 wt% fiber	195.28	16.87	367.27	78.97	4.17
10 wt% fiber	207.69	16.41	366.14	78.73	4.86
15 wt% fiber	199.11	16.74	365.98	79.16	4.10
20 wt% fiber	177.25	15.43	364.23	78.62	5.95

IOP Conf. Series: Materials Science and Engineering **947** (2020) 012010 doi:10.1088/1757-899X/947/1/012010

4. Conclusion

The mechanical and thermal properties of short salago fiber reinforced epoxy resin composites were evaluated in this study. Based from the results of mechanical testing, there have been improvements on the tensile modulus and impact strength of the composites. As compared to neat epoxy, the tensile modulus and the impact strength obtained for 20 wt.% salago fiber loading increased by about 83.5 % and 26.4 %, respectively. However, noticeable decrease in tensile and flexural strengths were obtained as the fiber content increases which may be due to the short fiber length and 2-D random orientation of fibers. Lastly, the initial peak temperature obtained from TGA increases from 5 wt.% to 10 wt.% fiber loading which indicates improvement on the thermal stability.

5. References

[1] Alzebdeh K I, Nassar M M A and Arunachalam R 2019 Effect of fabrication parameters on strength of natural fiber polypropylene composites: statistical assessment *Measurement* **146** 195–207

[2] Sanjay M R, Siengchin S, Parameswaranpillai J, Jawaid M, Pruncu C I and Khan A 2019 A comprehensive review of techniques for natural fibers as reinforcement in composites: preparation, processing and characterization *Carbohydr. Polym.* **207** 108–121

[3] Sanjay M R, Madhu P, Jawaid M, Senthamaraikannan S, Senthil S and Pradeep S 2018 Characterization and properties of natural fiber polymer composites: a comprehensive review *J. Clean. Prod.* **172** 566–581

[4] Balla V K, Kate K H, Satyavolu J, Singh P and Tadimeti J G D 2019 Additive manufacturing of natural fiber reinforced polymer composites: processing and prospects *Compos. B. Eng.* **174** 106956

[5] Kabir M M, Wang H, Lau K T and Cardona F 2012 Chemical treatments on plant-based natural fibre reinforced polymer composites: an overview *Compos. B. Eng.* **43** 2883–92

[6] Jeyapragash R, Srinivasan V and Sathiyamurthy S 2020 Mechanical properties of natural fiber/particulate reinforced epoxy composites – a review of the literature *Mater. Today Proc.* **22** 1223–7

[7] Pouriman M, Caparanga A R, Ebrahimi M and Dahresobh A 2017 Characterization of untreated and alkaline-treated salago fiber (genus wikstroemia spp.) *J. Nat. Fibers* **15** 296–307

[8] Pouriman M, Dahresobh A, Caparanga A R, Moradipour M and Mehrpooya M 2018 Morphological and physicomechanical analysis of high-density polyethylene filled with salago fiber *J. Appl. Polym. Sci.* **135** 46479

[9] Pouriman M, Dahresobh A, Moradipour M and Caparanga A R 2019 Thermal and nondestructive analysis of high density polyethylene filled with milled salago fiber *J. Appl. Polym. Sci.* **136** 47873

[10] Odusote J K and Oyewo A T 2016 Mechanical properties of pineapple leaf fibre reinforced polymer composites for application as prosthetic socket *Journal of Engineering and Technology* **7** 125–139

[11] Kerni L, Singh S, Patnaik A and Kumar N 2020 A review on natural fiber reinforced composites *Mater. Today Proc.*

[12] Eng C C, Ibrahim N A, Zainuddin N, Arrifin H and Yunus W M Z W 2014 Impact strength and flexural properties enhancement of methacrylate silane treated oil palm mesocarp fiber reinforced biodegradable hybrid composites *Sci. World J.* **2014** 1–8

[13] Saw S K, Purwar R, Nandy S, Ghose J and Sarkhel G 2013 Fabrication, characterization and evaluation of luffa cylindrica fiber reinforced epoxy composites *Bioresources* **8** 4805–25

[14] Chin S C, Tee K F, Tong F S, Ong H R and Gimbun J 2020 Thermal and mechanical properties of bamboo fiber reinforced composites *Mater. Today Commun.* **23** 100876

IOP Conf. Series: Materials Science and Engineering **947** (2020) 012011 doi:10.1088/1757-899X/947/1/012011

Thermal and Mechanical Joining of Basalt Fiber Reinforced Thermoplastic Laminates

A Al-Obaidi[1,*], P Rochala[1], M Kroll[1], T Hahn[2] and V Kräusel[1]

[1]Chemnitz University of Technology, Institute for Machine Tools and Production Processes, Professorship for Forming and Joining, 09107 Chemnitz, Germany
[2]Nordmetall GmbH –Materials and Impact Engineering, Adorfer Hauptstraße 16, 09221 Neukirchen/Adorf, Germany

amar.al-obaidi@mb.tu-chemnitz.de

Abstract. In this paper, quasi-isotropic basalt fiber reinforced polyamide PA6 was joined to aluminum sheets using three different joining technologies. The joints were manufactured by means of nut and bolt (NAB) as well as two-piece hollow riveting (TPHR) as current state of the art processes. Furthermore, additional joints were established by the new inductive contact joining (ICJ) technology. The obtained specimens were compared by means of tensile shear tests and by analysis of the fractured surfaces. It was found that the joints produced by ICJ are superior in comparison to joints manufactured by TPHR with regard to joint strength. In addition, during ICJ, the fibers of the polymer material are not damaged and the strength of the base material is maintained. Further optimization potential of the ICJ process was discovered during the investigations, so that potentially higher joint strengths as well as shorter process times can be expected, making the process interesting for future industrial applications.

1. Introduction

Due to their high strength, light weight and good corrosion resistance, fiber-reinforced polymers (FRP) are preferably used in the manufacturing of lightweight structures as for aerospace, automotive, sport equipment and consumer goods, machine construction, marine applications and civil engineering. Generally there are two types of FRP, depending on the polymer used as matrix for the composite structure: thermoplastic and thermosetting. As glass or carbon fiber reinforced polymers exhibit high tensile strength but also high costs, a composite that consists of low-cost fibers such as basalt and a thermoplastic matrix opens up new possibilities for the production of cost-efficient hybrid FRPs. In combination with light metal cover layers, the resulting material can exhibit high ductility and damage resistance [1]. Therefore, a joining technology for FRP and ductile metals like steel or aluminum is required. Many attempts were realized to join FRP with metal. Appropriate joining technologies can be classified into mechanical [2] [3], chemical and thermal processes [4] [5]. In this paper, a new process is presented to create basalt fiber reinforced polymer laminates (BFRP) by the combination of induction heating and the application of a joining force. The joints manufactured with this new technology were compared with mechanical joints of the same material.

2. Materials and experimental procedure

Basalt fiber reinforced polyamide PA6, a quasi-isotropic hybrid layer system with ten layers of basalt fiber reinforced PA6 (endless fibers) and unreinforced PA6 cover layers on top and bottom, was laminated with thin aluminum sheets of EN AW-5754 by means of two different types. The general preparation procedure of such laminates can be found in [1]. For the first lamination type ("Type I"), BFRP, was laminated between two aluminum sheets (figure 3) and for the second lamination type ("Type II"), one aluminum sheet was laminated between two BFRP layers (figure 3). In both cases, the thickness of the aluminum sheet was $d_{Al} = 0.3$ mm. In this paper, three methods for joining Al laminated BFRP with aluminum sheets are investigated and compared. The first one (ICJ) is a thermal joining method that uses induction heating. The two other methods are TPHR and NAB and thus are defined as mechanical joining methods. For the latter two joining methods, the aluminum sample and BFRP sheets were pierced and then joined. The TPHR samples were joined by means of a manual mechanical press. In the NAB joining process, the sheets were adjusted by M4 screws and the nuts were tightened together by a torque wrench with a torque of M = 3.5 Nm.

Joining by induction heating was utilized through the new hybrid process of ICJ by using an induction coil as heating, pressing and cooling device. Therefore, a water cooled induction coil is positioned on top of the overlapping parts and a defined joining force is applied. When the joining process is started by inducing an alternating current by a high frequency (HF) generator, the heat is generated in the electrically conductive material of the stack by induction heating. As heat is transferred to the electrically non-conductive plastic material in the adjacency of the inductively heated metal, the polyamide matrix melts and spreads over the metal surface. When the generator is switched off and the joining process ends, the joint is actively cooled by means of the water cooled induction coil. A detailed explanation of the ICJ process can be found in Kräusel et al. [6]. Figure 1 illustrates the ICJ process for joining of Type II BFRP and aluminum sheet in an overlapping configuration. The validation and evaluation of the investigated joining process was realized by tensile shear testing of the specimens according to DIN EN ISO 14273:2014 [7]. The testing principle can be seen in figure 2.

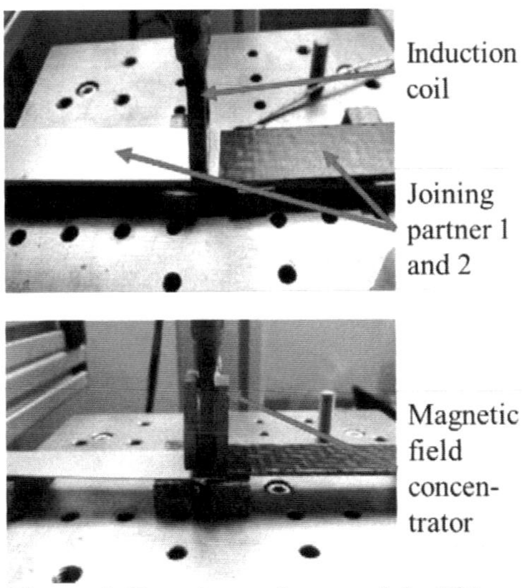

Figure 1. Experimental setup of the ICJ process.

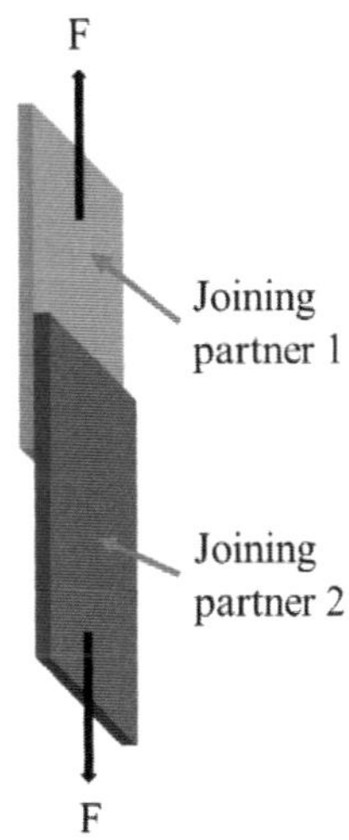

Figure 2. Principle of tensile shear testing.

For the joining experiments, aluminum sheets (aluminum alloy EN AW-5754) were used as supplied by the manufacturer ("Al/wop"). Moreover, the aluminum sheets were mechanically pretreated by sand-blasting ("Al/wp"). These aluminum sheets were joined to the two different types of BFRP mentioned before (Type I and Type II). The BFRP sheets had a thickness of $d_{BFRP} = 2.4$ mm and were cut into samples with dimensions of $A = 30.0 \times 160.0$ mm by water jet cutting. The overlap length was $l_o = 45.0$ mm according DIN EN ISO 14273:2014. The ICJ joining time was $t_{j1} = 3.0$ s, $t_{j2} = 4.0$ s and $t_{j3} = 11.0$ s for Type I + Type II, Type II + Type II and Type II + Al material combinations respectively. The generator (EMAG eldec Induction GmbH, Dornstetten, Germany) delivered a power output of P = 6.0 kW and operated at a frequency of f = 350.0 kHz. In Table 1, a matrix with all tested material combinations can be found. For each material combination and joining method at least 4 samples were manufactured and tested to ensure a minimum level of statistical certainty of the investigated joining processes. BFRP sheets of Type I and II that were joined by ICJ, NAB and TPHR are shown in figure 3.

Table 1. Tested material combinations for each joining method.

Joining method	ICJ	Two parts hollow rivet	Nut and bolt joining
Material combination	Type I + Type II	Type I + Type II	Type I + Type II
	Type II + Type II	Type II + Type II	Type II + Type II
	Type II + Al/wp	Type I + Type I	Type I + Type I
	Type II + Al/wop	Type II + Al/wop	Type II + Al/wop

For each of the three joining methods one sample was visually analyzed in order to identify the failure mode during mechanical testing. For the tensile shear tests, a universal tensile testing machine was used with a constant cross-head speed of $v_t = 0.1$ mm/s.

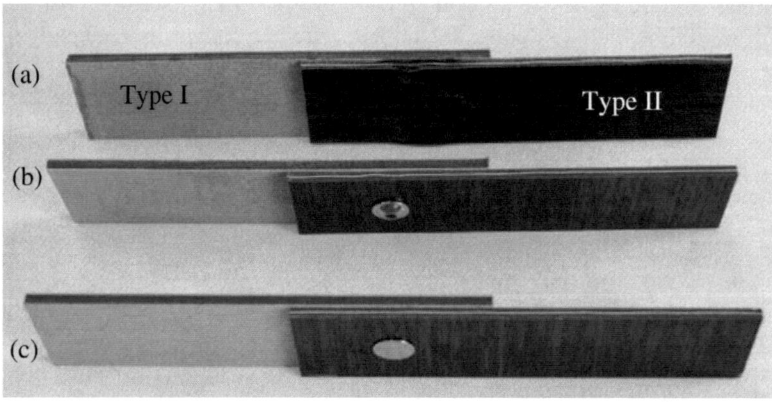

Figure 3. Samples of Type I and Type II, joined by ICJ (a), NAB (b) and TPHR (c).

3. Results and discussion

Figure 4 shows the mean values of the maximum tensile forces for various material combinations that could be determined for samples joined by means of ICJ, NAB and TPHR in the tensile shear tests. It can be seen that the NAB joints can bear the highest force before failure for all material combinations examined in this work. The measured tensile forces for NAB-joined specimens range from $F_{NAB\_min} = 4187$ N to $F_{NAB\_max} = 5368$ N. For the TPHR joints, only comparatively low tensile force values could be determined, which, however, are relatively constant across all examined material combinations. Thus, a value range of $F_{TPHR\_min} = 1107$ N and $F_{TPHR\_max} = 1247$ N could be determined. A more differentiated evaluation of the force values must be carried out with regard to the joints accomplished by means of ICJ. It becomes obvious that there are limitations concerning the choice of materials for the ICJ process as several material combinations cannot be joined with this technology. For example, sheets of Type I cannot be combined with each other and for the combination Type II + Al a mechanical surface pre-treatment of the aluminum sheet is necessary in order to realize a joint at all. The tensile forces of the joints produced via ICJ exceed the values achieved for TPHR joints. Maximum tensile forces of $F_{ICJ\_max} = 2803$ N were measured as well as minimum forces of $F_{ICJ\_min} = 1283$ N. It should be noted that the whole ICJ process was performed within joining times of $t_j \leq 11.0$s. As a result, the ICJ process can be regarded as a very efficient and productive joining process for multi-material composites.

Figure 5 shows the joints after the tensile shear test as well as their characteristic failure mode. For NAB and TPHR, figure 5 clearly shows that failure of the joints is primarily caused by ripped out bolts or fractured rivets respectively. Thus, the joining elements are the weak spots of the joint. This is because the perforation of the parts during the joining process causes damage to the fibers. Also the load bearing cross section of the laminates are reduced by the holes for the respective joining elements. As a result, premature failure of the base material is a likely mode of failure. With ICJ, the fibers in the sheet are not affected by the joining process, so that the strength of the base material remains unaffected. This is an advantage of the ICJ process, compared to the NAB and TPHR joining methods. By analyzing the fractured surfaces of the tested parts joined by ICJ, it can be observed that only a fraction of the overlapping surface has effectively formed a joint. The ICJ joining process can therefore be optimized in order to maximize the effective joining area, which is likely to increase the joint strength significantly.

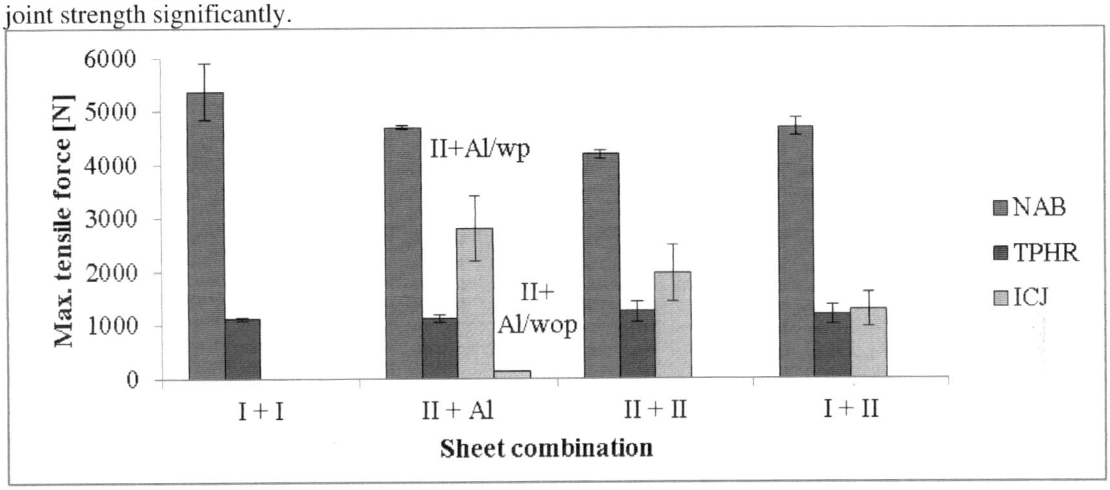

Figure 4. Mean value of the maximum tensile shear force and corresponding standard deviation of hybrid joints manufactured by NAB, TPHR and ICJ.

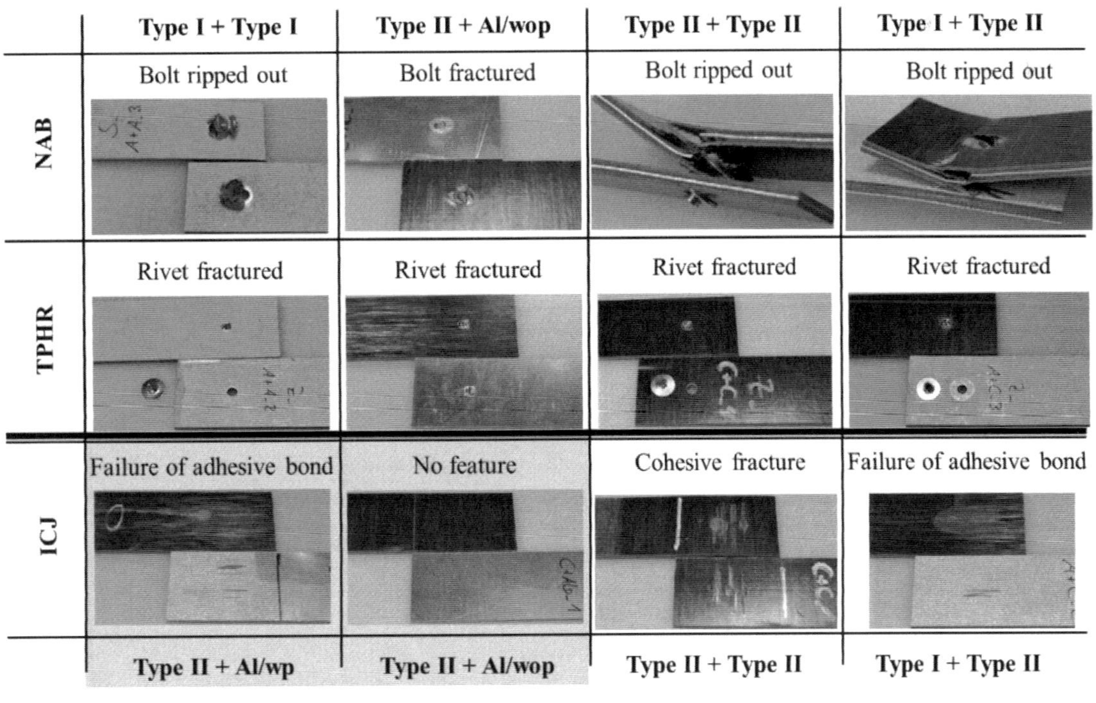

Figure 5. Summary of the fractured joints after the tensile shear test and description of their characteristic failure mode.

IOP Conf. Series: Materials Science and Engineering **947** (2020) 012011 doi:10.1088/1757-899X/947/1/012011

4. Conclusion

The ICJ process allows for joining of hybrid material composites within very short joining times. These joints are superior to joints manufactured by TPHR with regard to achievable joint strengths. However, the ICJ process still holds great potential for increasing the joint strength. Possible approaches are the maximization of the effective joining surface of the overlap as well as the mechanical or chemical surface pretreatment of both the metallic and the composite components. These approaches will be subject of future investigations.

Acknowledgments

The publication of this article was funded by Chemnitz University of Technology.

5. References

[1] Karapepas C, Trautmann M, Todt A, Al-Obaidi A, Nendel S, Kräusel V and Wagner G 2019 "Development of tailored hybrid laminates: Manufacturing of basalt fibre reinforced thermoplastic orthoses with aluminum thin sheets" *Key Eng. Mater.* **809** 245–252

[2] Lambiase F, Durante M and Di Ilio A 2016 "Fast joining of aluminum sheets with Glass Fiber Reinforced Polymer (GFRP) by mechanical clinching" *J. Mater. Process. Technol.* **236** 241–251

[3] Conte R, Buhl J, Ambrogio G and Bambach M 2018 "Joining of aluminum sheet and glass fiber reinforced polymer using extruded pins" *AIP Conf. Proc.* 1960

[4] Mitschang P, Velthuis R, Emrich S and Kopnarski M 2009 "Induction heated joining of aluminum and carbon fiber reinforced nylon 66" *J. Thermoplast. Compos. Mater.*, **22** 767–801

[5] Wagner G, Balle F, and Eifler D 2013 "Ultrasonic welding of aluminum alloys to fiber reinforced polymers" *Adv. Eng. Mater.* **15** 792–803

[6] Kräusel V, Fröhlich A, Kroll M, Rochala P, Kimme J and Wertheim R 2018 "A highly efficient hybrid inductive joining technology for metals and composites" *CIRP Ann.* **67** 5–8

[7] Normative, "DIN EN ISO 14273," 2014.

IOP Conf. Series: Materials Science and Engineering **947** (2020) 012012 doi:10.1088/1757-899X/947/1/012012

Kibble-Zurek Scaling with Matrix Product States

E M Howard[1,2]

[1]Department of Physics and Astronomy, Macquarie University
[2]Centre for Quantum Dynamics, Griffith University

eric.howard@mq.edu.au

Abstract. We study the dynamics of second-order phase transitions that may have taken place in the early universe and analyze the Kibble-Zurek (KZ) scaling as well as the formation of topological defects emerging from a quench in the one-dimensional Bose-Hubbard model. We describe the universal dynamics of the KZ mechanism (KZM) and the topological defect formation by simulating the Bose–Hubbard model at zero temperature. We study the effects of thermalization on the ground state by analyzing the transition between Mott and superfluid phases. The out-of-equilibrium dynamics induced by a quench of the hopping parameter is presented. We find a KZ scaling behaviour at zero temperature and analyze the KZ dynamics.

1. Introduction

In the early Universe after the Big Bang, cosmological phase transitions may have been generated by the expansion and cooling of the Universe. During this transitions, the symmetry of vacuum was broken and new vacuums within spacelike separated regions were formed. A challenging problem in current cosmology is whether the vacuum contains these topological defects forming during phase transitions in the early universe and occurring within 10−35s after the Big Bang in grand unified theories. After the cosmological phase transition at 1027K, the a phase of the vacuum is disordered, possibly giving rise to a number of topological defects. The cosmological mechanism behind these processes can be observed in condensed matter systems under certain conditions in laboratory. In condensed matter theory, when a physical system is rapidly driven through a second-order phase transition, topological defects (vortices, strings) are created at a density that is dependent on the speed of the underlying transition.

The KZ theory [1] predicts a variety of transformations from high- to low-temperature phases, associated with a spontaneous symmetry breaking and the evolution of topological defect structures like domain walls, strings, and monopoles. Cosmic strings may be directly related to the primordial density inhomogeneities that later gave rise to galaxy formation. These processes are also important for their connection to the grand unified theory (GUT) symmetry-breaking phase transition of the very early universe. GUT phase transition may have played a significant role in the inflationary dynamics of the early Universe. T. W. B. Kibble suggested [2] that such defect structures (domain walls, strings, and monopoles) may have appeared during the expansion and cooling of the early and very early universe. In recent years, most current experiments based on ion-crystals, spin-liquids, liquid 3He, liquid 4He, superconducting films, polariton superfluids, Josephson tunnel junctions and ultra cold atoms in optical lattices contain numerous findings with different levels of certaint which are consistent with existing quantitative KZM predictions [3]. KZM model [4] model has been applied to different scales from cosmological concepts to atomic systems, such as the Higgs field in the very early universe, superconductors or quantum fluids in condensed matter systems. The predicted behaviour at cosmological scale is present in laboratory when a system going through a second-order phase transition is cooled at a finite rate into the low symmetry phase. Cosmological inflationary theories have grown popularity mainly due to the lack of such defects within the visible cosmological horizon. We simulate the Kibble–Zurek mechanism [5] using Matrix product states simulations of the Bose-Hubbard model and discuss its nonequilibrium dynamics behaviour on the background of a cosmological evolution scenario. Within this picture, we investigate the nonequilibrium dynamics in a condensed matter analog of the inflationary dynamics of the Early Universe. The Kibble-Zurek mechanism [6] provides a description of the topological defect structure occurring during the symmetry breaking phase transitions, resulting in the presence of cosmological strings in the early universe or vortex lines in a superfulid. Specifically, ultracold atoms in optical lattices provide a unique setup to study the non-equilibrium quantum dynamics of defect structure formation. The recent progress in ultracold atoms physics provides key experiments for non-equilibrium dynamics and quantitative testable solutions for isolated, versatile and highly controllable systems. For example, a quench in pressure at a finite rate will drive liquid ^{4}He from a normal phase to a superfluid phase, leaving behind vortex lines, that may be associated with the existence of cosmic strings in the Early Universe. The defect structure is generated by the cooling processes when spatially separated domains are placed too far apart to exchange any information, due to finite communication speed allowed, resulting in multiple separated space-time regions with degenerated local broken symmetry [7]. The prediction of the formation of defects across a quantum critical point provides an intuitive theoretical framework for quasi-adiabatic quench processes across quantum phase transitions [8]. Kibble-Zurek hypothesis [9] provides a method to calculate the density of such defects when their symmetries are present, allowing to find the scaling laws for the density of defects as a function of the quenching rate through the phase transition and predict the growth of correlations and the density of defects following an asymptotically slow ramp.

We analyze the defect and domain structure quantitatively via Matrix Product States for Bose-Hubbard model [10] in a lattice regularised model. The mechanism describes the the nonequilibrium dynamics associated with the symmetries and the scaling laws after a phase transition, as long as its effects dominate the kinetics of the topological defect formation. The resulting dependence of the size of the causal domains where symmetry is overall broken is given by a power law with a small fractional exponent. The main prediction of this mechanism is the scaling of the defect density with the quench rate. During recent years, several experiments and numerical simulations studying the formation of defects in phase transitions induced by a quench both were carried out. Most systems of bosons in an optical lattice are well described by the Bose–Hubbard model, used in studying the superfluid to Mott insulator phase transition and realized in a multitude of experiments. The one-dimensional Bose–Hubbard model is particularly interesting because of the presence of a multicritical point with a Berezinskii–Kosterlitz–Thouless (BKT) transition. Findings from numerical simulations of the KZM scaling law [11] for exponents of the correlation length and the rate of topological defect formation with respect to the quench time are of direct relevance to the study of non-equilibrium behavior in cosmological scenarios. At the same time, the confirming experiments with ultra-cold atomic gases represent a step forward towards a better understanding of the broken symmetry vacuum, spontaneous formation of topological defects and the associated dynamical phase transitions in the early universe.

2. Topological defect formation and the Kibble-Zurek mechanism

The formation of topological defects predicted by field theory in cosmology lead to important consequences for the evolution of the Early Universe, associated to relativistic causality. KZM mechanism was mainly developed to predict the power-law scaling of the defect density across a continuous quantum phase transition. The adiabatic-impulse approximation tool approximates the time evolution of the quantum state of the system and has its origin in KZM theory applied to continuous quantum phase transitions. Spontaneous symmetry breaking assumes the degeneracy of the ground state of the system. When the system goes out of equilibrium, the broken symmetry phase contains different spatial regions defined by the different orientations of the broken symmetry. Therefore, in the new phase, independent choices of the vacuum will be associated with causally disconnected regions in the extended system. KZM scenario [12] is associated with nonequilibrium dynamics and second order phase transitions, as well as estimating the density of defects as a function of the quench rate through this transition. In quantum field theory, phase transitions are defined as transitions between one vacuum state and another. A quantum phase transition is a transition that is driven by quantum mechanical and not by thermal fluctuations. As a system approaches the absolute zero temperature, the thermal fluctuations are frozen out, and the quantum mechanical behaviour becomes relevant. If the Universe is in absolute ground state at a finite temperature, as it slowly cools, the ground state evolves continuously through what is known as a second order phase transition (or a continuous phase transition). If this change in the ground state of the quantum field is sudden and discontinuous, a first order phase transition occurs. When the change in temperature is slow compared with the relaxation time, the order parameter has to time to adjust to this change, and the evolution will be adiabatic. KZM predicts that the density of the topological defects in a second order phase transition should scale with the transition rate. The defects are mainly developing because of the divergence of the relaxation time near a critical point, resulting in a nonequilibrium dynamical process when the system is not able to follow the changes, causing the symmetry to be broken. During a non-equilibrium continuous phase transition, the evolution ceases to be adiabatic near a critical point as a result of a slowing down dynamics associated with the divergence of the relaxation time close to the critical point. When the system is smoothly driven from a disordered to an ordered phase through a second order phase transition, topological defects are formed and the scaling exponent of the defect density with the quench rate can be predicted by KZM.

We study the defect production upon slowly changing a system parameter. Let us assume that when crossing of the phase transition at a finite rate, the spontaneous symmetry breaking during a continuous second-order phase transition is controlled by a parameter λ. KZM will describe the dynamical evolution of a continuous phase transition under a time-dependent change of λ across the critical value. The evolution is characterized by the divergence of the correlation length ξ at equilibrium

$$\xi(\varepsilon) = \frac{\xi_0}{|\varepsilon|^{\nu}}, \tag{1}$$

as well as the relaxation time at equilibrium τ

$$\tau(\varepsilon) = \frac{\tau_0}{|\varepsilon|^{z\nu}}, \tag{2}$$

as a function of the distance to the critical point λ_c, the dynamic critical exponent z. and the correlation length critical exponent ν. The constants ξ_0 and τ_0 depend on the microphysics while ν and z depend only on the universality class of the transition. The notions of scale invariance and universality are central in the study of continuous phase transitions. Near the critical point, the characteristic length and the time scales diverge, generating collective phenomena decoupled from the microscopic degrees of freedom. The universality class of the transition obeys a power law in the quench rate, with the exponent given by a combination of the critical exponents of the transition. The same critical exponents will be associated with systems in the same universality class.

In this work, we consider a reduced distance parameter ε defined as

$$\varepsilon = \frac{\lambda_c - \lambda}{\lambda_c}, \tag{3}$$

As $\varepsilon(t)$ varies from $\varepsilon(t) < 0$ to $\varepsilon(t) > 0$, the KZM dynamics takes three stages: adiabatic, nearly frozen, and adiabatic again. When the system crosses the critical point, the phase transition will follow a spontaneous symmetry breaking throughout a change in dynamics from a high-symmetry phase defined by $\varepsilon < 0$) to a degenerate vacuum manifold for $\varepsilon > 0$.

If the system undergoes a linear quench symmetric around the critical point at $t = 0$

$$\lambda(t) = \lambda_c[1 - \varepsilon(t)] \tag{4}$$

the reduced parameter is dependent on the quench time τ_Q and varies linearly in time in $t \in [-\tau_Q, \tau_Q]$

$$\varepsilon(t) = \frac{t}{\tau_Q}. \tag{5}$$

When the equilibrium relaxation time is small enough and far from reaching the critical point $|\lambda|?\ \lambda_c$ following a quench (5), the system behaves adiabatic. Near $\varepsilon(t) = 0$, the divergence of the equilibrium relaxation time causes the system to slow down almost close to freezing, not being able unable to adjust to the changes of $\varepsilon(t)$. The dynamical evolution of the system does not stop and the order parameter will still continue to evolve locally, but ceasing to follow its global equilibrium value. Therefore, the local thermodynamic equilibrium of the microscopic degrees of freedom in the system can be still maintained. The evolution of the system slows down and stops being adiabatic when the phase transition is crossed at the critical point. The formation of topological defects is a main consequence of the local evolution of the Hamiltonian after crossing the critical point with a delay of $\hat{t}$ and setting up a boundary between the adiabatic and frozen stages. After the initial adiabatic stage, the frozen correlation length generates topological defects and thus the adiabaticity is broken. Consequently, any small change of the control or thermodynamic parameters will lead to non-equilibrium behavior.

The freeze-out time is given by

$$\hat{t} : \left(\tau_0 \tau_Q^{z\nu}\right)^{\frac{1}{1+z\nu}}, \tag{6}$$

The critical point fully controls the non-equilibrium behavior of the system, defining the system trajectory underlying the phase transition in the parameter space. Near the critical point, the equilibrium relaxation time diverges and it can be approximated with the time after crossing the critical point

$$\tau(t) \approx |\varepsilon / \dot{\varepsilon}| = t. \tag{7}$$

The freeze-out behaviour is due to the collective degree of freedom associated with the order parameter not keeping up with any externally imposed changes. In other words, the degrees of freedom associated with the broken symmetry will not follow any change of ε. This sets a delay between the order parameter and the equilibrium point associated with the parameter $\varepsilon \in [-\hat{\varepsilon}, \hat{\varepsilon}]$, with

$$\hat{\varepsilon} \equiv |\varepsilon(\hat{t})| : \left(\frac{\tau_0}{\tau_Q}\right)^{\frac{1}{1+z\nu}}. \tag{8}$$

The average size of the regions is defined by the equilibrium correlation length at $\hat{\varepsilon}$, following a power law of $\hat{t}$ when the relaxation time scales as a power law of ε.

$$\hat{\xi} \equiv \xi[\hat{\varepsilon}] = \xi_0 \left(\frac{\tau_Q}{\tau_0}\right)^{\frac{\nu}{1+z\nu}}. \tag{9}$$

If D and d are the dimensions of the space and of the defects, the density of topological defects can be calculated from the value of $\hat{\xi}$

$$n: \frac{\hat{\xi}^d}{\hat{\xi}^D} = \frac{1}{\xi_0^{D-d}} \left(\frac{\tau_0}{\tau_Q} \right)^{(D-d)\frac{\nu}{1+z\nu}} . \tag{10}$$

3. Results

In the following, we investigate the main features of the system's state and its dynamical behavior after a quench and verify the KZM scaling mechanism for zero temperature, by employing powerful numerical tools such as Matrix Product States and Tree Tensor Networks. We analyze the out-of-equilibrium dynamics scenario in the one-dimensional Bose–Hubbard model at zero temperature. We use the Time-Evolving Block Decimation (TEBD) algorithm and use a Suzuki-Trotter decomposition of the Hamiltonian at second order. More specifically, the system is undergoing a linear-ramp quench in the hopping strength J across the second order quantum phase transition, starting from zero temperature. The time evolution of the of the quantum many-body state is here analyzed, starting from the equilibrium state ρ_0 of the Hamiltonian H, which is time-dependent through a linear ramp in the hopping strength

$$J(t) = \frac{2J_c}{\tau_Q} t + J_c , \tag{11}$$

where $J = 0$, and $\mu = 1/2$. If we compare the system's internal relaxation timescale $\tau_R(t)$ with the external driving timescale $\tau_D(t)$, the dynamics splits into two stages: an adiabatic stage for $\tau_R(t) < \tau_D(t)$, and a sudden stage for $\tau_R(t) > \tau_D(t)$. The system's dynamics changes from adiabatic to sudden during the "freeze-out time" $\hat{t}$. KZM scaling is present if the properties of the system after the quench are dependent on the instantaneous ground state at $\hat{J} = J(\hat{t})$.

Nevertheless, the Hamiltonian is symmetric around the critical point J_c with $J(0) = J_c$ and $J(-\tau_Q/2) = 0$. As the total particle number N is a constant of motion, we therefore have $[H(t), \sum_{j=1}^{L} n_j] = 0$. Specifically, at zero temperature, ρ_0 is the pure state composed of the perfect Mott insulator state $|\Psi\rangle$ with filling one, i.e. $|\Psi\rangle = |1\rangle_1 \ldots |1\rangle_L$, while at finite temperature $\rho_0 = e^{-\beta H_0} / \mathrm{Tr}[e^{-\beta H_0}]$. The initial state ρ_0 is evolved via unitary time evolution $\dot{\rho} = -i[H(t), \rho]$ in the time interval $t \in [-\tau_Q/2, \tau_Q/2]$ and remains a product state at all times. Here, τ_Q is the quench time, while the coupling term vanishes.

Asdiscussed above, the KZM mechanism predicts that the scaling of the final density of defects as a function of the quench time time is determined by a single constant exponent κ. The final correlation length will therefore scale as

$$\xi_{\text{fin}} \propto \tau_Q^{\kappa}, \quad \text{where} \quad \kappa = \frac{\nu}{1+z\nu}, \tag{12}$$

if at the critical point the equilibrium correlation length diverges with a critical exponent ν and the energy gap ΔE closes with another critical exponent $z\nu$. Analitically, the final correlation takes the form

$$\xi_{\text{fin}}(\tau_Q) = 2\sqrt{J_c}\,\tau_Q + \mathrm{O}(\tau_Q^2). \tag{13}$$

The exponential scaling of $\xi(J)$ at equilibrium, near the critical point J_c can be approached by fitting the exponentials to the equilibrium quantities with a power-law around around the freeze-out point $\hat{J}$, resulting into the "effective" critical exponents ν_{eff} and $[z\nu]_{\text{eff}}$ The scaling from Eq. (12) is now recovered by replacing the exponent κ with an effective exponent $\kappa(\tau_Q)$ at zero temperature:

$$\kappa(\tau_Q) = \frac{\nu_{\text{eff}}(\tau_Q)}{1+[z\nu]_{\text{eff}}(\tau_Q)}. \tag{14}$$

that depends on $\hat{J}$ and on the quench time τ_Q. We numerically compute the freeze-out times $\hat{t}(\tau_Q)$ and the effective critical exponents ν_{eff} and $[z\nu]_{\text{eff}}$, finding that a KZM scaling of the final defect density is here present. For different values of the quench times τ_Q, three different regimes are distinguished: a sudden quench regime for τ_Q ˆ 2, where the system is driven by a short impulse of duration τ_Q, the KZM scaling regime, for 2 垐τ_Q τ_U (where τ_U is the upper bound for τ_Q and it is due to the finite system size) and a finite-size saturation regime when the system is completely ordered and the final correlation length is saturated due to the finite system size. The KZM scaling of the final defect density occurs after the quench time $\tau_Q \geq 2$, when the driving timescale is not sufficiently fast, as for the case $\tau_Q < 2$, so that the quench will begin to slow down and allow the relaxation timescale to catch up with the changes in the system.

4. Conclusion

In summary, using Tensor Network methods as a simulation tool, we studied the effects induced by the initial conditions in a driving critical dynamics and confirm the validity of KZM mechanism for a quench of a quantum system at zero temperature in the one-dimensional Bose–Hubbard model. Furthermore, by analyzing the scaling of the final correlation length ξ_{fin} after the quench and the nonadiabatic behavior of the "defect measure" as a function of the quench duration τ_Q, we find a non-trivial KZM scaling behaviour, in agreement with the theory.

5. References

[1] Kibble, T. W. J. 1976 *Phys. A: Math. Gen.* **9**, 1387.
[2] Kibble, T. W. 1980 *Phys. Rep.* **67**, 183.
[3] Zurek, W. H. 1993 *Acta Phys. Pol.* B **24**, 1301.
[4] Zurek, W. H. 1996 *Phys. Rep.* **276**, 177.
[5] Zurek, W. H. 1985 *Nature* **317**, 505.
[6] Zurek, W. H, Dorner U, Zoller P, 2005 *Phys. Rev. Lett.* **95**, 105701.
[7] Dziarmaga J, Rams M. M. 2010 *New J. Phys.* **12**, 055007.
[8] Dziarmaga J, Zurek W. H 2014 *Sci. Rep.* **4**, 5950.
[9] Gardas B, Dziarmaga J, Zurek W. H 2017 *Phys. Rev. B* **95**, 104306.
[10] Shimizu K, Kuno Y, Hirano T, Ichinose I, 2018 *Phys. Rev. A* **97**, 033626.
[11] Zurek, W. H., Bettencourt, L. M. A., Dziarmaga, J, Antunes, N. D. Shards 2000 *Acta Phys. Pol.* B **31**, 2937.
[12] Laguna, P. & Zurek 1997 W. H. *Phys. Rev. Lett.* **78**, 2519.

AUTHOR INDEX

Institute of Physics
Dirac House, Temple Back
Bristol BS1 6BE UK

ISSN: 1757-8981
ISBN 978-1-7138-3169-3
9 781713 831693